The Oort Cloud

The Oort Cloud

A Discussion of Some Important Topics of our Times

By

Marvin Dixon

ISBN: 1983652105
ISBN 13: 9781983652103
Library of Congress Control Number: 2018900789
CreateSpace Independent Publishing Platform, North Charleston, SC

Preface:

I wrote this book in the form of an essay as a way to leave behind some of my thoughts on some of the big and important topics for my grandchildren and anyone else that might be struggling with large questions for which there are few satisfactory answers and even less verifiable proof. For instance, questions about the origin and subsequent demise of the universe, the purpose of life, man's place within the universe, religion and the concept of God, intelligence, and space travel. My approach was to talk each of the topics from my own personal perspective as a result of my own reasoning. In many cases, you'll find that my thoughts are generally consistent with classic cosmology but in others, there is a considerable amount of divergence from current thinking. And in some areas my reasoning is completely foreign and inconsistent with the endless theories that have already gained some level of societal as well as scientific acceptance; specifically, my position on a possible universal life force, intelligence being a requirement for life, dimensional space and dark matter and dark energy, as well as a few other areas. But it is important to keep in mind that what is presented here are only my positions and my beliefs and while well-reasoned, you will find that only 2 of my positions can be supported by facts or proof; everything else is open to further debate, modification, or outright rejection.

It is my belief that when we struggle to find answers to some of the large questions we move man one step closer to becoming a higher functioning being and intellectually enlightened which, by my thinking, helps in his moral enrichment. And while we are currently prisoners of our Oort cloud, it is my hope that by the time we actually become interstellar voyagers, we would have better answers to many of the large and important questions that continue to elude us. With that knowledge, it would be my expectation that we would have grown as a species and would be better able to decide the morally difficult challenges were apt to encounter during the voyage and at the point of disembarkation.

What follows are my thoughts from inside our Oort cloud. I hope you'll find this work thought-provoking.

Marvin Dixon

Dedication:

I dedicate this book to my grandchildren so they will have a better understanding of what their grandfather was really thinking with respect to some of the big and important questions of today and throughout time. I also dedicate this book to those scientist and philosophers that wrestle with these questions that may not have real answers or have answers that are beyond our current level of comprehension. And it is also dedicated to those that can imagine how the impossible would become possible at higher dimensions of space-time.

Marvin Dixon

Contents:

The Oort cloud

A Discussion of Some Important Topics of our Times

The Oort cloud:

The Oort cloud is a theoretical bubble that is believed to form an envelope around our solar system at a great distance from our sun. It was named for the Dutch astronomer Jan Oort who predicted the solar system was surrounded by trillions of icy bodies (planetesimals and comets) that were under the gravitational influence of our sun. Oort's prediction was an attempt to explain erratic and long period comets; comets that have never been seen or recorded, and comets coming into the inner solar system from directions other than the Kuiper belt; directions other than the planetary plain. We live inside the cloud and its outer boundaries mark the limits of our solar system. For example, the outer boundaries of our solar system are at an average distance of 2.2 light years from the Sun. In order to put this in to some visual context, I'd like for you to imagine putting a marble on the floor and let it represent our sun. Then, put a single grain of sand one inch away from the marble to represent the distance of the earth from the Sun. Now put another grain of sand 40 inches away from the marble, we will let it represent the orbit of Pluto. Pluto's orbit would encompass the orbits of all the planets in the inner solar system. Finally, if we want to really get an appreciation of the true distance of the outer boundaries of the Oort cloud we will need the use of a car. Using our scaled down model of our solar system, the outer boundaries of the Oort cloud would be more than 2 miles away from our marble. The cloud is vast, more than 4.4 light years across with little more that empty space between our star and our nearest neighbor; Proxima Centauri which is about 4.3 light years from our sun. If we boarded a spacecraft and headed to Proxima Centauri at the speed of Voyager 1; 10.6 miles per second, it would take us over 37,500 years to just reach the outer boundaries of the Oort cloud, and over another 37,500 years to reach Proxima Centauri's inner planets. And even then, when we got there we might find that nobody was home.

Of the eight planets in our solar system (acknowledging the demotion of Pluto) we live on the only one that is known to harbor life. We are alone in our cloud and currently lack the technology to even colonize Mars, which is far less challenging than it would be to move beyond our solar system. Now I'd like to change your point of reference, with respect to the Oort cloud from the very large to the very small. We live in the Milky Way galaxy among over 200 billion other stars; all with their own Oort clouds and potential planetary systems. And since Edwin Hubble expanded our perception of the size of our universe, astronomers using the telescope named in his honor, have determined that our universe is made up of more than 400 billion other galaxies. They've calculated the universe to be over 27.4 billion light years across; our Oort cloud is infinitesimal by comparison. We are adrift and helpless in a very dangerous galaxy that wouldn't care if our little star blinked out and in a much larger universe that wouldn't even notice if our entire galaxy were to disappear. But here, on earth we have convinced ourselves that we're special and somehow protected or immune from the forces of nature; but we're really not. This, near limitless universe was not created for the benefit of mankind. If it were, it would not have been so impossibly large and continuing to expand at ever increasing rates. And our universe is a very dangerous place. As we look out from our little cloud, we are at constant risk of comet and asteroid collisions, wandering black holes, rogue planets, gamma ray burst, solar eruptions, and the pollution of the planet as well as being in danger of self-annihilation. Any of these events could trigger a mass extinction and could even extinguish the entire planetary gene pool. Earth would become just another barren rock adrift in space. If we truly hope to be able to move off in to our galaxy and begin seeding it with humanoid lifeforms then that should cause us to start thinking about the universe, our place in it, and the natural forces that we are subordinate to in new and different ways.

Scientists all over the planet are currently hard at work theorizing and designing advance propulsion and life support systems in order for us to move out and explore our solar system first hand. But in order to get beyond the Oort cloud into interstellar space will require us to develop new science and that will take time. And while I have little doubt that we will develop and continue to improve technological advances I can imagine none that would get us beyond our local group of stars. Even if we could travel at light speed it would still take 120,000 years to cross from one side of our galaxy to the other. But attaining light speed is a "catch-22" and our current math puts it beyond our reach, because with an increase in velocity there is a proportional increase in mass which requires a greater input of energy. Eventually you'll reach a limiting point, a point where acceleration stops because infinite mass resist infinite energy. So unless and until we can find ways to bend time and space, interstellar travel would be slow to develop and intergalactic travel impossible.

If the current science is accurate and verifiable then it brings into question some of our fundamental beliefs, particularly our perceived place in nature and our relationship with our divinity. For example, who and what are we, what is our purpose, and why are we here? Why is there life and what is its purpose and benefit to nature? Is life unique or a product of naturally occurring physical and chemical forces? What and where is God?

I don't think that life on earth is unique or an anomaly. I believe (without benefit of objective evidence) that there is a *"Life Force"* that permeates throughout every region of the universe and much of that is intelligent life. But because we have not been able to find any signs of intelligent life by listening to the cosmos, we have to assume that other intelligent life must be at some great distance from us. This then means that we are affectively alone in our small neighborhood and on our own. If we ever got in to trouble, there would be no place we could go, and there'll be no one to come to our rescue. Our survival is dependent on new and revolutionary thinking and the actions we take to preserve earth's genetic diversity and heritage, now and in the future. And while I am not suggesting some rapidly approaching doomsday, we should not just assume that time is on our side either, because it's not. Any of the events noted could occur without giving us enough time to defend against them or to evacuate the planet[1]. And at the time of writing this, the probability of greatest risk to the planet and the survival of mankind is man's unrelenting pollution of the planet and the real potential for self-annihilation.

---

[1] The total evacuation of over 7 billion people from the planet is a logistical impossibility. At best it would be possible to evacuate a critical mass of human diversity (genetic heritage) and a supporting ecosystem and bio fauna and flora to support them.

Qualification and Disclaimer:

I think it appropriate to state from the outset that I don't have any special information or insight in to life, the workings of the universe, the nature of time and space, god, or the concept of reality (those things that are real and obvious, those that are real and imperceptible, or those things that are simply imagined). We can only look out from our tiny rock in our tiny cloud and make observations and use our science in order to draw some basic conclusions about the things we see or perceive. But we are limited in what we can see and perceive. For example, without the use of instruments, we are blind to the entire light spectrum except the tiny little part where visible light is emitted. We are limited in how we perceive the world around us and can only perceive time in seconds and can have no real perception of our universe expressing itself in *Planck Time*, *Deep Time* (cosmological time), or time unfolding at other rates. Additionally, we can only see and perceive our world in 3 dimensions and with the passage of time (time being a separate dimension that is independent from any of the other spacial dimensions). Super string theorists predict the existence of an additional 7 spacial dimensions over the 3 that we know. But we are blind to them, we are limited to 5 senses and we are limited as to what we can perceive with them. We can't do remote viewing, mental telepathy, telekinesis, teleportation, or sense dark matter or dark energy (reported to make up over 95% of the energy and matter in the universe). And when we look out from our little rock and make predictions about other distant rocks we run a significant risk of being wrong. So therefore, what follows are my personal thoughts, postulations and my beliefs. None of which is based upon immutable facts, only supposition from observations, personal logic and my best guess. And, more likely than not, what I am supposing is also likely to be wrong. But here goes…

The Universe:

The universe's existence is measured in deep time, over billions of years, and by comparison, we have only begun to study the heavens less than a second ago. So when we make predictions about the origin of the universe we'll have to leave room for a considerable amount of error. If we leave aside the mythical and religious creation accounts, we are left with 2 popular models: The Big Bang Theory and The Steady State Theory (also call The Continuous Creation Theory).

In the Big Bang model the universe exploded into existence from nothing and is still expanding. This model suggests that the universe will continue to expand until every particle of mass is pulled apart by dark energy and evaporates. They call this *The Big Freeze* or *The Big Whimper*. But some Big Bang theorist argue that there may be enough matter in the universe to slow its expansion enough to allow gravity to take hold and cause the universe to contract back to its original point of origin. They call this possibility *The Big Crunch*.

The Steady State model fell out of favor in the 1960s as the Big Bang model replaced it. It proposes that the universe is uniform from every direction and had no beginning and has no end. That though it is expanding, its density remains constant due to matter and energy being constantly created. I am not sure if the steady state model is consistent with another model I've heard of where the new energy and matter comes from outside the universe. That model suggest that though the galaxies are moving away from each other, the density of the universe remains constant as a result of continuous creation from outside the universe. For example, when matter and energy collapses into a black hole in a neighboring universe, the matter and energy is cutoff from that universe and is transferred to ours via a "white hole"; a mini big bang. The transferred energy and matter condenses into new galaxies thus maintaining the density of our universe. We know that there are super massive black holes at the centers of all galaxies and we know that there are stellar black holes roaming our galaxy but we have yet to document the existence of a white hole.

If the continuous creation model is correct, there should be some evidence of them, but we've seen none. This particular concept has been taken further in support of the *Multiverse Theory*. That is, that there is not just one universe but an untold number of them forming from collapsing black holes. For example, the Multiverse Theory suggests that our universe is the product of a single black hole collapse in a neighboring universe creating a white hole which in turn became our entire universe.

Another possibility for the Steady State model could include an initial big bang, where the explosion pushed apart a force that was confining the infant universe and allowed the runaway inflation of dark energy. This would have created an outward expanding bubble containing the universe as an internal bubble of dark energy pushed against it. The expansion of dark energy and retreat of the confining force would continue until the outward and inward forces reached equilibrium and became stable and static; thus, Steady State. In this model, the universe would exist as a thin membrane on the outside of a bubble of dark energy and the inside of a bubble of opposing force. And, the bubbles could be smooth and spherical or reticulated like the folds on the human brain.

Before sharing my own thoughts, I would like to just mention a relatively novel possibility. It might just be possible that we are living in a very ancient universe and from our perspective, it is boundless. It could very well be *The Mother of All Universes,* and ours is only one of trillions that exist within it. Our universe could have come into existence as a result of a *super hyper nova* of one of the stars in that universe. What we've determine to be the Big Bang that brought our universe into existence may have only been the results of an exploding star. And, what we've imagined to be the universe and the galaxies around us might only be star and galaxy formation as a result of gravitational accretion of the dusty and gaseous nebula that resulted from that explosion. Such an explosion would have surely left behind an unimaginably large gravity well, a black hole in the center of our universe. When scientists have looked across the universe, they've determined that the expansion of the universe was speeding-up when they've theorized it should have been slowing down. In view of the contradiction between the theoretical and the observed, scientists have attributed the phenomena to *Dark Energy* working against the universe's gravity that would normally be retarding the universe's expansion. If we were to set dark energy aside, because dark energy and dark matter can't be measured, could there be other forces at work? For example, could a massive black hole at the center of our universe be twisting and folding space to make it look as though space was expanding when it was in fact contracting? Specifically, could we be observing the curvature of space from the surface of a structure that is similar to a *Real Projective Plane* or a *Klein bottle,* and what we see is an optical illusion caused by the warping and folding of that space? Or, could there be a gravitational force adjacent to our universe (in the larger universe) that is drawing the outermost stars and galaxies in our universe toward it at an ever accelerating rate? Just a few possibilities, but for now, we have no way of knowing.

Moving on to my thoughts… My thoughts with respect to the creation of the universe are a bit different but are more in line with the Big Bang Theory. In my model, the energy in the universe is finite and will be conserved so there will be a big crunch before another big bang. And in order for you to visualize the concept it would be helpful if you got yourself a strip of paper and some tape to make a Mobius strip. I suggest the Mobius strip because time (as we know it) only moves forward and events in space change with time. At the point where you taped the two ends of the paper together we will label the topside "MG&MP" for maximum gravity and maximum pressure and on the bottom side we will label it "MDE&mp" for maximum dark energy and minimal pressure. As you move along the Mobius strip from the instant after MG&MP, you would be moving along a tiny membrane just after the big bang and the creation of the universe. As you move further along in time, the universe expands due to the explosive nature of the big bang and inflation. Gravity begins to weaken the further we move away from the instant of creation, and dark energy begins to gain strength. As the temperature in the early universe begins to cool it goes through a series of changes, energy is converted into matter, clouds of matter condense in to the massive stars, the stars explode in super nova creating the more complex elements, second and third tier stars and their planets are created and begin to form galaxies. Dark energy continues to force the galaxies and space apart as gravity continues to weaken[2]. Bright stars will use up their energy and go dark leaving only white and brown dwarf stars, neutron stars and black holes. Eventually all the matter in the universe (both visible and dark matter including black holes) will be converted into dark energy and gravity will be at its weakest and dark energy at its maximum. It is at this point where those that believe in the *Big Freeze* scenario and I part ways. The belief that the universe (both time and space) will continue indefinitely sustained by only a handful of electrons and positrons that might find one another and react in the infinite vastness of space or until they and the universe completely evaporates is difficult for me to accept. If the universe evaporates, then just where does it go, and what about the conservation of energy? Their scenario would require an open system that is devoid of energy leaving only static time and space. I don't understand how time and space could be supported without energy; they would have no other means of sustaining themselves and should collapse out of existence.

The events that we have walked through until this point would be considered the evolution of the universe, from birth until death. For our purposes, they are generally consistent with the current thinking on the model. So for visualization and starting at MG&MP on your Mobius strip you can mark out the events as they would have occurred. For example, the primordial era, the star and galaxy formation era, the expansion and degeneration of the galaxies era, the dark matter and black holes conversion to dark energy era, and the dark era where all matter has been converted to dark energy. By my thinking, this would take us half way around the strip bringing us to MDE&mp, where dark energy is at its maximum and the expansion pressures are at a minimum. At this point, there will be no more matter in the universe that can be converted into energy. At that instant, gravity

---

[2] Quantum Mechanics and Super String Theorist suggest that gravities' weakness in the third dimension is due to its force is being stretched across all 10 of the dimensions that they have predicted.

would be almost nonexistent, space would stop expanding, time would stop, and the temperature of the universe would be at, or near, zero degrees Kelvin.

Now, in order for my model to work, there would have to be a conservation of energy, and that would only be possible if we lived in a closed universe, and I believe that to be true. If the universe exploded or expanded from nothing into existence, it had to overcome some force that kept it confined. Therefore it had to expand from nothing into something. And while I have no idea as to what that something was, it had to form a bubble around the expanding universe as it was being overcome by the initial force of the explosion and the later forces exerted by dark energy. When dark energy had no more matter to convert it would have lost its momentum and stopped its push against the bubble. At that point, the bubble would have begun to contract and to reclaim the space it lost due to the energetic expansion cause by the big bang. And with respect to the conservation of energy, the energy that was used to expand the bubble during the expansion of the universe was returned as a function of the elastic properties (force) on the bubble.

As the bubble contracted, it would have begun to concentrate the latent heat of the universe, increasing its heat energy and increasing the forces of gravity and pressure. The bubble, in concert with the increasing forces of gravity, would have increased its rate of collapse exponentially until temperatures increased to trillions of degrees. As the universe would get smaller the temperatures would grow increasingly hotter and the forces of gravity and pressure would have become irresistible. At the instant just prior to its collapse, the universe would have become infinitely small and infinitely hot. If you refer back to your Mobius strip, we would now be at 1 second before MG&MP and just over a second before the *adiabatic explosion* that triggered the big bang[3]. In this model, not only was energy conserved, but time and space as well. Time continued to move forward though at an ever decreasing rate. As the density of the contracting universe increased, the

---

[3] In thermodynamics, an adiabatic process is one that occurs without transfer of heat or matter between a thermodynamic system and its surroundings. In an adiabatic process, energy is transferred to its surroundings only as work. In this case, the work is the re-expansion of the universe.

universe's gravity and internal pressures also increased and time would have slowed down until it nearly came to a stop. And because space did not tear during the expansion of the universe, it would continue to contract containing all of the energy in the universe as well as time; exactly the way it was prior to the big bang.

In my model, as with most of the others, the assumption is that the universe is finite. But if the universe is infinite, all the models would fall apart.

Also, in my model, I believe that the universe might exist along a single or multiple membranes with space and time being its fabric. And because there may be multiple membranes, there may also be multiple possibilities for different universes and different Mobius strip scenarios. Multiple universes could be the result of imperceptible energy fluctuations at the instant where gravity collapses the universe and causes time to move along a different random path, along a different Mobius strip and a different possible universe. There may be no time or space between the universes, only energy, or there may be only a small gap between the possible universes in order for time or space to be attracted to one, or make the jump from one to another. This would be the only time where a particular universe could be selected and the variations could be infinite; the second between universe collapse and rebound. Once a universe had been selected, the other possible universes would no longer be viable because they would not have existed in *real time and real space* and gravity would have collapsed them out of existence. Therefore, the forces of gravity collapsing one universe rebounds at some point (no one knows where, so I'll just call it a *singularity*) to provide the energy needed to activate time and open space for the next…the next Big Bang.

It is important to note that neither my model nor any of the others have been able to explain the beginning, the origin of the forces that put the universe into motion; creation itself. And as to why the universe exists, and what is our role in it? You can find any number of soothsayers, mystics, and diviners that will tell you whatever you want to hear, but I have absolutely no idea. So far, I haven't seen any role for us on the universe's stage above observer status, and even then, we are stuck way up in the cheap seats and have really little understanding of the drama that is unfolding before us. This is not to say that we don't have a role, it's just that it's not clear to me what it could be. But for now, we have only been observers peeking through a hole in the cosmic fence watching a split second of the billions of years of development in the life of the universe; and we really don't understand much of what we've seen. Therefore, as spectators, we're left with far more questions than answers.

Now, this is where things get confusing...all of the above models (including my own) spoke to specific theories describing the creation of a three dimensional universe; and they all sound reasonable at the outset. But because there is a prediction that there could be as many as ten different dimensions and one of time, in that arena, all the models fell apart or became inadequate when I created my model of a ten dimensional universe and attempted to theorize the concept of dark energy and dark matter, the expansion of the universe and intellect. But that model bares very little resemblance to the one I proposed above. In that model, time and space expands the universe in order to accommodate increasingly higher spacial dimensions and dark matter and dark energy are spacial dimensions that cannot be viewed at a lower dimensional state; they are invisible to those existing in the third dimension but apparent to those existing in the tenth. You will find the specific contrast much further down in the section entitled *"The Grand Design...the Multi-dimensional Universe"*.

The Universal Life Force:

There is some form of life everywhere you look on the surface of our planet, in our oceans, and in greatest abundance in the soils and beneath the surface of the waters. Life is everywhere and I don't think it's by chance, but a product of the physical and chemical laws that were established at the instant of creation. Space, time, and energy are the immortals over the life of the universe. They are forever changing but are fundamental to the universes existence. The next step down are what the ancients called the first elements; earth, wind, fire, and ice. They were born as gravity condensed clouds of hydrogen and helium during star formation and the subsequent violent destruction of the first stars when they exploded in super novae. Physics gave birth to the elements, created the minerals, and the conditions necessary for chemistry to begin. And with chemistry came the initial precursors for life. Because star creation and destruction was ubiquitous throughout the early universe, it would follow that physics and chemistry were equally ubiquitous. As a consequence, we could assume the precursors for life and the potential for life would be equally ubiquitous throughout every region of the universe as well.

Life appears to be transient and dependent upon chemistry and the environmental conditions (environmental physics), those that happen to be favorable to its initiation and proliferation. Therefore, the type of life that sparks and evolves is subject to the chemistry that is possible in that physical environment. When conditions change, life either evolves to conform to the changes or that form of life dies out. But life is tenacious, and though a lifeform or lifeforms may die out, the precursors for life will endure and they will continue to endure until new and favorable conditions allow new and possibly different chemistry to spark the creation of new and different lifeforms. So, if the chemistry for life is distributed throughout the universe and is only waiting for favorable environmental conditions, the question becomes, is life a result of chemical happenstance or is it an actual force within the universe? *Could the universe be alive?* Could the universe be an all intelligent, living and sentient entity that exists everywhere in every state (living and nonliving)? My short answer is... *"I don't know"*. But even more baffling to me is "what is the purpose of *our* lives, and what is its value and benefit to the greater universe"? Is life as we know it, an inevitable artifact of the 4 forces that govern the universe (the strong force, the weak force, electromagnetism, and gravity)? Or as some have suggested, life might be a parasite. That due to environmental happenstance and favorable conditions, life is little more than an opportunistic parasitic organic blight hitching a ride on the back of an inorganic universe. I take issue with their assertion because, by definition, parasites come in from outside the host in order to exploit a host. So, if life were a parasite, then just where outside of the universe did life come from? Life is a product of the universe itself so that would make any talk of a parasitical relationship off the mark and ridiculous. I also find it difficult to imagine life in terms of chemical happenstance. Because the precursors for life (if I had assumed correctly) would be evenly distributed throughout the universe it would suggest that there is a force that initially programed life from the start and continues to sustain it in forms we may not yet be able to imagine; it insists upon life wherever environmental conditions allow. But the question still remains... ***"why"***.., what is life and why does it exist in a universe that apparently doesn't need it and appears to be totally indifferent to it? Again, I don't have any answers… I just don't know.

Going forward, I will leave aside any discussions of how the precursors for life were formed and how inorganic molecules sparked organic life, how life on earth evolved along 2 tracts (*chemosynthetically* and *photosynthetically*), and, as to whether life on earth got its start as a result of *Abiogenesis* or *Panspermia*. There is no real agreement on these theories and they are well beyond my intended scope for this essay as well as beyond my comprehension. But I believe that it was here that the miracle of life occurred, at the instant where photo-electric energy and the energy stored in inorganic matter was converted into organic life by chemistry and physics. It was at this point that life became a force within the universe. But we still don't really know what life or the life force truly is, aside from a general description that we've constructed to explain what we think we know about what is living and what is not. My questions go a bit deeper; for example, does life exist as *centillions* upon *centillions* of separate and discrete unrelated living entities or is life connected to all other life in some way by some grand unifying force, or does life exist as a combination of both; connected one to another as well as discrete points of existence?

Without proof or any evidence to support my position, I choose to believe that there is a force throughout the universe that drives and sustains life everywhere. In addition, I believe that though individual life is undeniably finite, the life force within the universe is eternal. These are only my beliefs; nothing more than speculation and conjecture on my part. So as I describe what I believe to be the force that connects and supports all life, please keep in mind that my probability of being *"wrong"* is all but assured.

I believe that all life is connected and the force that connects and controls it might operate in a similar fashion as does the water cycle on earth. Though, unlike the earthbound water, the universal life force is only limited by the limits of the universe. Therefore, by my thinking, there is only one water on the planet and that water (regardless of its form; liquid, solid, or vaper) is connected and will change forms and locations (from oceans, rivers and streams, ground water and underground aquifers, clouds and ice), but it is still only one water. It is discrete in raindrops and diffused in clouds but connected in rivers, lakes, the oceans and ice. And because I believe that there is only one water, I also believe that there is only one life force... *"The Creator, **God**"*.

If we continue to use earth's water cycle as the model, it would suggest that there is really no death in the life force, only transformation and relocation. For example, diffuse life energy floating in the clouds, unrelated and undefined. They condense around some unknown particle and fall from the sky as *"new"* and unique and discrete living entities similar to rain drops or snowflakes. Upon reaching the ground, they are absorbed into the soils and into the root systems of plants bringing life sustaining water to the earth's increasingly more complex life forms. All of these life forms, whether bacterium, animal, plant or insect can be considered discrete individual and unique living entities. And they use chemosynthesis, photosynthesis, and predation in order to continue to concentrate the energy of the life force. From that point forward the life force is further concentrated in the biological food chain until the host dies. When the host dies, some of the concentrated life force is again recycled and some returns to the streams, rivers, and oceans…back to the *"ancient"* body of life; *the life force.* It is my contention that the life force is a phenomenon and should be quantifiable on some level and by some method we have yet to discover. As gravity acts upon everything in the universe discretely and individually as well as collectively and simultaneously across the universe's great expanse, I believe that we'll find that the life force acts in the same way sparking and sustaining life wherever environmental conditions allow. If the life force is found to be made-up of particles then *Quantum Mechanics* could explain how life particles interact and are entangled over great distances. But with respect to death, *"the end of existence"*, I believe that if death were *"the end of existence"* and an actual reality in nature, then life could not and would not be self-sustaining and it would have no purpose… none what's so ever. And it is on this point that I part ways with the *Existentialist* because I do believe that the universe and life has a purpose, though I lack the intelligence to understand what it might be. We tend to be narrowly focused with respect to our existence and what it means to be alive. We measure existence in terms of consciousness, intellect, and the passage of time. We choose to only measure our existence as either living or dead, when they could very well be two sides of the same coin. And if we exist, then it is quite possible that we have always existed simultaneously or alternately in both states, living and dead; or better, sentient and self-aware or unresponsive and oblivious to the passage of time. The atoms that make up our bodies will continue to exist throughout time as will the energy that gave us life. So, we can approximate when we became sentient but not when we began life.

Lastly, life may very well imply intelligence, and the physical laws of the universe may be operating by default or by *sentient intelligence*, though without life as we have so narrowly defined it. So when I refer to a life force (in this instance) I am speaking of a biological physical force, a bruit and persistent one with a single focus…and that is just to survive. But life may very well exist in ways we cannot even imagine and with a broader definition could include intelligence. And if so, intelligence may prove to be the universes greatest mystery. Could intelligence *be proof of God or is God intelligence?* If not, than where does intelligence come from? Is it a product of the brain or a function of universal mathematics and logic? Is intelligence self-limiting in man or is it limitless? And how can we measure universal intelligence when we only have man's intellect as our highest point of reference? These are just a few of the many questions for which I have no answers and that leaves me troubled. I fear that we might find that we are in the early stages of our evolution and are at a much lower level of intelligence than we may have thought. Because our solar system is relatively young (only about 4.5 billion years) there are likely to be much older civilizations in older solar systems that are relatively close-by (in a cosmological sense) that could be billions of years more advanced and far more intelligent than we are. For example, when we measure man's intelligence we arrive at a *mean intelligence value* (IQ) of 100. What would be the outcome if we encountered a civilization that would measure 200, 300, or even more using our methods? If that were the case, and if contact were ever made, we might find ourselves at their mercy and would have to rely on their generosity and benevolence for our very survival. The opposite is also a problem… we could find civilizations or living beings that are less advanced than us; having lower intelligence and without an ability to resist or defend against us. If that were the case, what would we do? How generous and benevolent would our predatory nature allow us to be with them? Will we be able to resist exploiting and enslaving them for our selfish purposes? Will their civilizations and future development even matter to us or will we just brush them aside in an act of planetary genocide? We don't have a good track record here because we've abused, enslaved, and exploited before and we're doing it right now. Far worse, would we cultivate them like cattle, pigs, chickens, fish, etc., and just use them as another food source? We might do just that, and because turnabout is fair play, another more advanced species might just do the same thing to us. And if they were significantly further advanced and intelligent, would they or could they even recognize us as being an intelligent life form at all? Would we even matter to them? I'll leave this as an open point for discussion because this poses a genuine risk to all of mankind and it is something that deserves further and sober consideration should contact ever be made.

We are sentient, insightful, intelligent, animated and aware of the passage of time, and by our own definitions, that makes us alive and living. Some force is acting upon us to make that happen and you are free to call it whatever you choose, *The Universal Life Force, God, or anything else.* But I cannot believe that our existence is due to happenstance because *"basic life"* in its many diverse forms, as well as intelligence appears to me (without getting into a debate over *Evolution or Intelligent Design*) to be too intricate to be a product of chance. I believe that there must be an underlying force that is omnipresent throughout the universe, and that force forces life everywhere.

Let's move on to God and Man's Religions…

The Existence of God:

Man seems to have always been blind to the apparent, those things that are well within his senses and glaring. We don't see anything of value around us but instinctively stand on our toes and look off in the distance for those things that are less than at arm's length. So when it comes to finding God, are we looking here and there and everywhere we can to find God and missing the real possibility that God is manifested in us and in everything we can and cannot perceive? We want and even expect our God to be an identifiable entity that is concentrated in a fixed place that we can point to and say there [he] is, sitting on [his] throne in the heavens and conducting the affairs of men on earth. All the religions of the world have their own unique beliefs and stories as to how and why God created the earth and the heavens. But alongside those stories and deeply held beliefs were those that questioned the very existence of God. Then there were others that questioned whether "God created Man" or did "Man create God". If those questions were put to me, I would have to say "Yes"; I believe all three to be true. And to find the answers to each we would only have to look inward, deep within ourselves.

To the first question, "Does God exist"? Based upon my limited ability to perceive and interpret the world around me and what little grasp I have of elementary logic, I would have to answer "yes", God does exist. If what I perceive in the physical world is real, if those creations are beyond the capabilities of men to create, than the creator of the earth and the heavens must be what we consider to be God. My position is based upon personal observations and primitive logic, not faith.

And with respect to the second question, "did God create man". If we consider the question as objectively as we can we should be able to agree upon the *fact* that "we exist". It doesn't matter how or in what form of existence we believe; whether we exist in a physical state, or are a part of some cosmic dream, a bunch of "ones and zeros" in a binary computer program or as Rene Descartes put it, I think therefor I am, or if we exist as a result of any of the religious creation stories. In some form or fashion we exist and some force or entity caused that to happen; God. And in the absence of direct actions of a named all-powerful god, theoretical physicists suggest that we might still exist. They would argue that the universe and everything in it could have sprung into existence by happenstance and from nothing at all. According to physicist, the *laws of physics* do not prohibit the universe from occurring spontaneously, without a trigger and from nothing at all. This could help to explain the "Big Bang Theory"; *something or everything from nothing at all*. I can't argue the math nor rationalize the concept of something from nothing. But this is only one of many concepts we can't rationalize, for example, the beginning of time, the end of time and eternity. What existed before the beginning of time, what will exist after the end of time, and eternity is time without beginning or end. Therefore, if it is our understanding that the universe could have sprung from nothing at all, to me, it would represent an overt admission that we lack the ability to understand the concept of ***"nothing"***. The closest we can get to nothing is the absence of something. But in place of something that is absent, there is still empty space and time which is still something. That then means that we haven't gotten any closer to identifying what nothing is. So, if the universe and all that's in it sprang into being from nothing at all, then that nothing must be God. And by extension, because we cannot define or conceive nothing, we cannot define or conceive God; the creator of man and all that there ever was, all that there is, and all that will ever be, either.

And as to whether man created God, again I would have to say "yes". The creation of the world's gods and the belief in them is self-pacifying and anchored in "faith"; man's ability to believe what he cannot see or understand. And this is in direct follow along to my statement above. Because we cannot define or conceive God, we have created our own gods and have made them identifiable, relatable and tangible to our understanding; we've made our gods in our own image. From the very earliest glimmers of man's intelligence we've recognized that there were forces and powers well beyond ours and our abilities to control, and they filled us with fear. They became our first pantheon of gods: the god of the Sun, of the moon, of the earth and forest, thunder & lightning, of earthquakes, disease

& death, etc. Subsequently, we've created many more pantheons of gods to help to explain and protect us from the causes and effects of human and natural events in our world. And the relationships that we've formed with those gods can be distilled down to their base constituents. While we refer to them as religions or religious practices, when you get down to the meat of it, they are nothing more than *grand bargains*. For example, we glorify our gods and pay tribute to them in the form of adoration, veneration and exultation so long as they favor us and protect us from harm and misfortune. We've offered up gifts and payments to our gods in the form of food and water, blood, animal and human sacrifice, gold and untold riches, and we have built temples and monuments in their names. During times of turbulence (drought and famine, sickness and death, etc.) we've concluded that the gods must be angry with us and require special tribute and ceremonial rituals in order for us to regain their favor. Or, we would come up with new and different gods, sometimes to replace old ones and at other times to explain misfortune. For example some of the gods we've created are malevolent and their anger explains extremes in the weather, earthquakes, erupting volcanoes, crop failures, plague, war, death, etc. And for the individual, we've created lesser gods or demons to explain our baser acts (malice, rage, aggression, greed, vengeance, and insanity), or any other actions that can be considered an internal injury to man's humanity (man's soul). As a consequence, we believe our gods afford us divine protection and will keep us safe and in favor, and keep the demons and darker gods away, so long as we continue to worship and pay tribute to them. This bargain, this agreement, this contract is the basis of our many religions and creeds within religions and faiths. Furthermore, because most of us find ourselves compelled to believe what we are told and eager to accept an authoritative explanation of any overt contradictions in our religions we rarely question the religious dogma or make any attempt to think for ourselves. We cower and are intimidated by the cult of religion and acquiesce to the current interpretations and just *"believe what we are told "*. We don't seem to have any historic religious memory aside from folklore and myths and have forgotten that the gods we worship are the very ones that we have created.

So in essence, we've entered contracts with our manmade gods, and have become prisoners of our faiths and trapped by our own imaginations, ignorance, superstitions, and have become afraid of the metaphysical gods of our minds.

So "yes", I fully believe that based upon the objective information I perceive and understand to be real about the earth and the heavens, "God does exist". And because man also exist, in some form and by some actions of a creator (whether as a result of any of the religious creation stories or from nothing at all) that creator must be God. Hence, I believe "God created man". But because God is undefinable and inconceivable to man, man has created the earth gods to stand as proxies to the God of the universe in order to provide him parental comfort and security in an unpredictable, frightening, and dangerous world. In following this line of thinking, I would have to say, that for purely selfish reasons, man created the many earth Gods. So therefore, it would be fair to say that "man created God" *(metaphorically)* as well as God created man *(empirically)*.

The Concept of Religion:

Religion and religious ritual is what we do to honor, praise, and glorify our gods. It is what we do to pacify and manipulate them so they become obliged to serve us and do our bidding. We offer them what we believe they want and intern, we expect that they will keep us safe from harm and always keep us in favor. And while our choices of gods and establishments of religious ritual may have originated out of ignorance, superstition and fear, they have evolved into widely excepted and unquestionable truths and sincere beliefs. For the believer, religion is no longer a selfish agreement or contract with a divinity, but a commitment to serve a higher being for a higher purpose than one's own. And to that end, nothing is ever questioned because no voice rings louder and truer than the voice of their god(s). And this becomes the fundamental problem with any religion in a world where there is more than one. One group has little-to-no tolerance for the existence of the other. The intolerance is not just limited to differences between religions but it extends to different sects within a religion and different creeds within religious sects. The disciples of one religion or sect hold their beliefs so tightly and passionately that they refuse to consider and outright reject any thinking that is not consistent and compatible with their own. They take advantage of any and every opportunity to promote their faith and enlist converts while simultaneously discrediting any others. And the danger here is that, for the most part, this is not done out of malice but out of love and a belief in their own god and a genuine consideration for those they see as pagans, heretics, and non-believers that are without god or, for those that have made an investment in some other deity that they consider to be, a *false god*. Religions are political in nature and are ultimately designed to control and subordinate a people. The god (or gods) and the religious rituals are chosen and developed by the group's leaders who've made a significant investment in their personal acquisition of political power and wealth and the pacification and ultimate control of the masses. To that end, they've developed gods that were consistent with superstition, myths, and rituals that were intended to protect man from the wrath of these gods. They knew that their religion would have to be intuitive to the followers and plausible and compelling to potential converts before it could take hold and become widespread. In addition they knew that there had to be a significant element of risk *(political intimidation and even physical harm)* for anyone that tried to resists their religious doctrine. For example, there could be retribution from the gods or a non-believer could be alienated from the group, or could be singled out as a heretic. In most cases, the cult of religion (religious intimidation) would be persuasive and seductive enough to encourage the majority of the group's members to want to be a part and to believe. But if not there are religious enforcers in every religious group; the zealots. The zealots make it their mission to convert the non-believers (by force if necessary) and to stamp out and destroy every vestige of their false gods and hedonistic rituals. But with their total belief in their own god(s) and the writings in their holy books, religion takes some zealots down a more extreme path where some have accepted the killing of those they cannot convert or will not convert to their religion as a service to their god(s). They become the soldiers of their god and justify killing in [his] name. And when this occurs, more often than not, the religious leaders remain silent. And in the minds of the zealots, that silence infers complicity and approval.

I have to go a little deeper here because there is at least 2 areas that are glaring and problematic for me when I consider the mindset of the zealots; the first is that the zealots appear to see their god as being weak and apparently impotent in the presence of man, and the second is how can one determined which zealot is killing in the name of the *"True God"* and which is terribly misguided and is killing in the name of a false god. Translation, which killer has god on their side? The former presents a conundrum and an inability to understand that there is a contradiction in their religious text, and the latter is a selfish and hypocritical act of violence that's filled with passion and devoid of logic and simple reasoning; only self-serving animalistic aggression.

With respect to the first…all religions proclaim their god to be omnipotent, the creator of all that is, omnipresent and with unlimited intelligence. Therefore, in my view, it would follow that an all-powerful god would have no need for religious thugs and enforcers. If such a god wanted man to bow down and worship [him][he] would have ordained it and man would have no other choice but to comply. In addition, I contend that if that were the intent of god, god would not have created man with the gift of free-will and the ability to choose. The zealots apparently see things differently. While they accept their god as the all-powerful alpha & omega and the creator of all things on earth and in the heavens, they believe that their god's power is weak and limited when it comes to getting man to see the *"True Path"*; the path of *their* particular religion. Their actions are supported by their religious text which encourages its followers to expand their doctrine by enlisting additional converts. The zealots interpret the text as a direct decree from God giving them the responsibility to corral and shepherd as many as possible into the light of their god. And they reject any notion of man's free-will because they consider man to be blind until he has been made to see the *"True Path"*, as proscribed by their religious teaching and their god. They eagerly enlist into the armies of *their* god in order to force the non-believers to accept their god, and non-other. And with just as much passion, they silence any voices of dissent, so non-believers should keep quiet or they should be prepared to die.

And to the second point…you would have to assume that if 2 different religions or religious sects were at odds and if their differences descended into war (a holy war) because each is advancing their own religious doctrine and, if you also assume that god wants you to kill in [his] name, one would be following the true god and one a false god. So logically, and at the very best, one would be right and one wrong. How could you determine which is which? From my perspective, both are woefully misguided. Let's leave behind the zealots and move on…

In view of our baser nature, I am hard pressed to see how man could have developed large, stable and relatively peaceful moral societies without religion and the fear of the gods. While it is my contention that morality is inborn in man, without the perceived oversight of man's divinities and their religious teachings, man's sense of morality may have been feeble and may have only extended to his family and immediate tribal members. And without the belief that the gods were always vigilant and were constantly looking over his shoulder; judging him and keeping records, man's morality might have withered each time he was confronted with a choice to do what was right or to opt-out and do what was expedient and low risk. Religion and societal norms within a religion has forced us to more thoroughly examine our actions based upon those standards and allowed us to extend our sense of moral obligation and our moral codes beyond our immediate family and tribe to the larger society. Because of man's warlike and predatory nature, it is my assertion that early civilizations may have never arisen were it not for a belief in a dynastic divinity. Subsequently, one could argue that our greater sense of morality and much of the progress we've made toward building and sustaining large civilizations can be traced directly back to the creation and proliferation of our religions.

So if religion played a role in the socialization of man and the development and expansion of his morality, it must then be a good thing. If we took a look at the world's great religions, I think we would have to say, in general and at face value, they've all been a benefit to mankind. They've provided order and stability in a world that may otherwise have been chaotic. They've put man in touch with higher beings and given him a set of ethical and moral codes with which to live by, and provided hope for man's greatest fear; *the fear of death.* But our view of religion has always been general and at face value. Few of us have ever bothered to look beneath the surface or to question glaring inconsistencies in the religious text and rituals. Our view has been decidedly myopic because none of us would want to question our religious leaders, go counter to popular beliefs, or far worse, offend the gods. I have long questioned those inconsistencies and have not been able to find satisfaction for them and their contradictions in the explanations given by religious leaders that avoid the encounter by avoiding the questions, only saying "god works in mysterious ways". Or they would reject any notion of contradictions or inconsistencies and would reduce it to a matter of interpretation; and obviously, from their perspective, my interpretations were wrong. I may very well be misinterpreting the apparent, but what I perceive to be apparent is inconsistent and contradictory. And to me, what is *apparent* is also *reality,* and no amount of authoritative, magical and mystical statements will make it otherwise. I will speak to some of the contradictions and inconsistencies further down, but for now, I want to talk a bit about man's soul.

If you are a religious believer and if that religion teaches that man has a "soul", and you believe that to be true, you are obligated to learn the truth about your chosen religion in order to protect and preserve your own soul. I can't speak to other religions, religious sects, or creeds within a sect; only my own. I believe that I am real and that I exist somewhere within this vast universe. I also believe that I am made up of the same materials that make up the universe and that I am a discrete, self-aware and sentient being that is a witness to the universe itself. Therefore, I believe that I have intelligence and have free-will that governs my actions. I will call this intelligence and free-will my soul. And while there is an enormous possibility that I have missed the mark here; this is what I believe. In follow along to my logic, if I have a soul, then everyone must have a soul. And like mine, theirs is as discrete and unique to them as mine is to me. Religions divide man's being between the physical body and the ethereal soul and because the natural deterioration and decomposition of the body after death is apparent, they promise that the souls of the believers will be preserved in Paradise or in a higher form of existence, Nirvana, and thereby man can avoid the horrors of Hell. This is the *"Great Hope"* of man and the promise of religions, though there is no proof that the promise has ever been kept. So man believes and lives and dies on hope. I have not been able to separate man's physical body from his soul because intelligence and free-will is totally dependent on a normally functioning physical brain. But if you believe that you exist and were created by The Creator (the God of the Universe) than I believe you have a soul and have always existed, in some form, and will continue to exist, in some form, until the end of the universe.

Man's relationship with the gods is at the center of all religions, and each religion charts what man must do in order to gain the favor of the gods, avoiding the wrath of Hell and reaching paradise; the salvation of the soul. Their position is that man is inherently flawed and without proper guidance will compromise his innate morals and place his soul in danger of damnation. Others start man off at a deficit. They teach that because of the acts of the first parents (Adam and Eve) man was born with original sin and the only way they could redeem their soul was through the actions of their savior and by following the teachings of their religious leaders. But if we would step back and take a more critical look at the world's religions we'd see that they all have been setup and operate as corporations that caters to man's ignorance, superstitions, and fears. This is not a criticism of religions because I believe that they play a valuable role in man's life by doing good works and by providing both spiritual and emotional support, but their organizational structure is a matter of fact. They are organized in a rigid hierarchical configuration where those at the top exercise the most power and influence, and share disproportionally in the wealth; and those further down benefit in accordance to their status and rank. Their business plan is targeted at the acquisition and maintenance of political power and garnering wealth to support their mission; *to dominate and rule over the masses*. Their wealth and political authority comes from their followers, and their task is to assure the perpetual growth and expansion of their faith, and theirs alone. Therefore, in a world where man has a choice of religions, he has to decide which provides the level of comfort, reassurance; intellectual inspiration and spiritual guidance he prefers to protect and preserve his soul. But religious freedom is relatively new. For many years there was no religious freedom because, in many cases, the leader of the religion was also the tribal leader, the king, or the head of state, and there were no choices. You either adopted the religion of the political authority or you were persecuted, exiled, or even executed. Even in today's world, many countries have official state religions, and while many of them tout religious tolerance, there really isn't very much in the way of tolerance at all, but there is no end to religious conflict and war when religions collide. When you look beneath the surface of today's religious conflicts, take a moment to try to determine their *actual root causes*. Are they being caused by actual differences in religions, religious interpretations, or are the true causes more related to the acquisition of **power**; specifically economic power and political dominance of the few over the many? What is your thinking? Let's just move on.

Religions also teach that there is a special covenant between man and the gods, that regardless of the many thousands of lifeforms on earth, only man is worthy of special consideration because god made man in [his] own image. Some speak of the beast of burden and the animals of the forest, but only as a backdrop to the stories of creation. Some others have assigned mystical powers to some animals and have made them spirit guides. And some use animals as temporary vessels for the soul to denote the journey one must travel during the many incarnations of the self before they reach enlightenment. But none has suggested that any of the animals possessed a soul independent of man's.

I've often found that through our own ignorance and arrogance we go through life labeling things we really don't understand. But once we've given something a label we mistakenly believe that its label defines it and gives us some degree of understanding about it; but it doesn't. Take the notion of the soul for instance. We all have some sense of what we believe the soul is. And religions mark a clear boundary between man's physical body and his immortal soul, but no one has been able to point to man's soul and clearly define what it is and how it differs from the body. Although, they can say with absolute certainty that man is the only being on earth with a soul. How can we be sure? We created the gods of the earth and we made us special to them, so in turn, we had them create us with a soul. But The Creator, The God of the Universe, created all that is in it; are we also special to that god? Did that god designate man to be the only living being on earth special enough to have a soul? And in all the vastness of the universe, and all the possibilities of other life forms, are we that special and alone in having a soul? I can't speak for other lifeforms, but I know that I exist and I believe that I have a soul! And that should be the end of it…but it's not. My assertions are based upon my beliefs, so *I believe*, and I am confident in and hold very tightly to those beliefs. But belief without proof is *faith* because I can't prove any of my assertions. I, just like you, have beliefs and faith, but have **"proof of nothing!"** And it is at this point where religious teachers confuse the weak with the following conundrum; they say "if you are a believer, no proof is needed, but if you are a non-believer, no amount of proof is enough". The weak are expected to acquiesce and believe what they are told, and they are told to have faith; **blind faith**, and they do. The religious leaders and faithful's point to their holy books and scriptures as proof of their divinity because they are considered to be the actual *"written word of god"*. But all religious text and rituals were authored by men, possibly very well meaning and inspired men but men nonetheless; not god. And most religious followers are unable to make that clear distinction, so they believe and believe with passion. Because of that passionate and narrow-minded view, I have made a personal policy to avoid all discussions of religion. People having the same religion, within the same sect and creed still have different interpretations as to what they believe, and even finer and obscure points of differences will lead to disagreements and verbal if not physical conflicts. The question becomes, "are they discussing and defending what they truly believe or only what they have been told to believe"? It's really hard to tell, but I've found no value in those discussions because they lead nowhere. And to my thinking, the religious texts don't provide a definitive and plausible definition of man's soul either.

When it comes to prayer, giving thanks and devotion, I have had a reasonable share of good fortune in my life and have benefited from the kindness and generosity of others at different points along the way. On each occasion, I have thanked the people or the fates that were responsible for my good fortune or contributed to it. In most cases the people were known to me and I thanked them directly, and I attributed my good fortune to the fates to my *Earth God* because that's as close as I can come to the *God of the Universe*. My expressions of gratitude, in each case, were sincere and heartfelt though of short duration; I thanked them and moved on. I have also suffered a number of personal and financial setbacks that threatened to devastate my world and me along with it. During those times I prayed and asked for divine intervention. But my prayers were not directed to the Creator of the Universe, but to my Earth God. Intellectually I know that each of the many gods of earth stand as representations of the Creator and were created by men, but I can't conceive of the Creator and during times of great stress, I run and hide in a safe and familiar port (behind my Earth God), and it provides me comfort. My devotion to my Earth God can be attributed to a combination of ignorance, superstition, and fear. But when I consider prayer and devotion and giving thanks to the gods as prescribed by many religions (including my own), I find that what is prescribed is inconsistent with devotion to any god but more in line with what would be prescribed to show devotion to a prince, a king, a dictator or a tribal or political leader; in essence, what one would do to show their devotion to a mortal man. And that's where I have problems.

Religions prescribe and define how its followers are to honor, praise, and glorify their gods through their writings and rituals. They inform man of the agreement between him and his gods and what he must do to reach the heavenly paradise, eternal life, and the salvation of his soul. Conversely, they warn man that if he fails to follow their teachings (the dictates of the gods), he risk their wrath, the loss of paradise, the damnation of the soul, and eternal sorrow in the pits of hell (or an equally unpleasant fate). So it follows that if man is to realize the promise of the gods, he must sing the praises of their names, offer up tribute and spread their word. My problem with this is that these are the stated requirements of the Earth Gods, not the requirements of the Creator. The Creator is silent and has made no requirements for a need for adoration or for tribute. Therefore, religion of any type was never a requirement of God, but men. And because it is my belief that we created the Earth Gods, it follows that it was we that also authored the religious writings and created the religious protocols and rituals. If you were to do an objective examination of the writings and practices of the world's religions you'd find that they are all steeped in magic and mysticism that is designed to terrify the non-believer and provide hope to those who believe provided they submit and subordinate themselves to the faith. And the keepers of the faith are the organized religions; hence the keepers of the faith are men. And it is men, not God, that profit from religions.

Kings, Caliphs, Maharajahs, Pharaohs, Potentates and Dictators all expected and demanded that their subjects adore and pay tribute to them in forms as varied as professing their greatness and exalting their name, pledging their allegiance to them above any allegiance to the state or to any god, bestowing gifts of gold and other riches, contributing a percentage of the harvest, and pledging their lives and the lives of their children during times of war, etc.,. From my perspective, the world's religions make the very same demands on their faithful. And should any one resist, the cult of religion will intimidate and demoralize the follower until they comply or are ousted from the community or worse; no different than being an enemy of the state. The need for adoration and tribute is a need of man, not a need of god. If you have belief in any of the creation stories and you believe that the gods made man, than Gods greatness is apparent and does not need to be stated or celebrated; it is what it is. The Gods of the Earth and the Creator of the Universe aren't vain or in need of reassurance in their position relative to man's. They didn't create man for the single purpose of having him prostrate himself and grovel at the foot of their altars praising their names. So they don't require man's assessment and verdict to be contented and self-assured in their own omnipotence. But it is a requirement of men and many of the world's religions. And when we pray and prostrate ourselves, are we really paying tribute to our gods or are we trying to manipulate them so our souls will be allowed to survive our bodies after death? Because many of us pray many times a day, it is important for us to truly understand why, why we pray. What are we actually doing when we pray? Are we really glorifying and giving thanks to our gods or are we selfishly and hypocritically attempting to deceive and placate them for our own benefit (for the salvation of our own soul)? And if it is the latter, who are we really deceiving?

Another manmade construct of religion is the notion of a heaven and a hell; the carrot and the stick. Once again, the Creator remains silent on any such notions of heaven and hell but religions force the earth gods to inform man of both and make *hope* and *fear* the centerpieces of their agreement with man. I have had hopes and fears like everyone else and the promise of heaven and the fear of hell used to top the list. But now I find that there are just too many inconsistencies in the religious text, mythical writings, moral inequities, and basic logic that have brought this notion, this basic agreement with the gods into question for me. Religions offer the promise of soul salvation to all that follow their teachings. But some appear to be favored over others. For instance, access to heaven is all but assured to those that consider themselves *"The Chosen People"* and for those that follow a specific god. Followers of other gods and religious sects are excluded and considered pagans. People that have lived a life of excess, wickedness, and debauchery are allowed to enter heaven provided they accept an earth god prior to their death. Though, a person that lived a life of righteousness can be denied entry due to some offence they have yet to atone for. The soldiers of god that are martyred while killing in [his] name or during the defense of the state have their souls instantly sped away to paradise or taken away to Valhalla by the Valkyries. Yet some religions deny entry to heaven of newborns that died without the benefit of baptism. The wealthy relatives of the dead can buy indulgences in order to redeem their souls and allow them to be freed from purgatory and ascend in to paradise. And the Pharaohs and noblemen of Egypt were given special knowledge from The Book of the Dead that guided them through the underworld. The poor and those without living relatives to petition the gods on their behalf had no champions or guides. Without champions and guides they had little hope of making it to the other side. And what of women, servants, and slaves what is their position in the hierarchy of heaven; what are their prospects and what is their role?

At this point, it should not be surprising that I can no longer believe in a heavenly paradise just flowing with milk and honey, a place where all my dreams and wishes come true. But if I was selling a faith, this is probably where I would start. I have always doubted the existence of hell due to the teaching and contradictions in my faith. I was taught that I had a loving and merciful god, and I believed it. Therefore, in my mind, such a god would never have created such a place to torcher and torment man for all eternity. What crime could man have possibly have committed in his short lifespan that would have warranted such a sanction? And conversely what good could man have possibly have contributed over the span of his lifetime to deserve a place in the heavens among the angel's and gods? Though, without a belief in man's perception of heaven or hell I am not without hope, or in fear of the loss of my soul. I exist and am self-aware and I believe that qualifies as having a soul. I am in the universe and I am of it. Scientist predicted the universe came into existence around 13.7 billion years ago (1.37×10^{10} years) and our star was born about 4.5 billion years ago (4.5×10^9 years). And they predict that the very last molecules in the universe will evaporate and the universe will disappear in an age that is measured by a *vigintillion* of years; in 1.0×10^{93} years. That's a #1 followed by 93 zeros. I have no memory of the birth of the universe or of our sun. But because I am in the universe and of it, I existed in some form during both events even though I was not sentient and self-aware. And when the last molecules evaporate at some great time in the future, the last of me will evaporate along with them. But for now and over the last 70 years I've been a witness to the existence of the universe itself, and that alone makes my existence spectacular and makes me privileged. I have no insight as to what will happen to my soul at the time of my death, but I am tempted to believe that there may be a life force that permeates the universe and manifest itself in every living thing. I believe that individual life radiates from it and then returns. The cycle of life and death continues but the body of life always remains constant. If this is so, then that force would be the Creator," God". And because we are living in the creation, we must also be living in God. As a consequence, after death, my soul will follow the path that the Creator has ordained; it will have no other choice, and I, no objection. It's also possible that life is simply a discrete chemical and biological event where life is sparked; it runs its course and then ends, If that happens to be the story of life, than it would be according to its design and beyond my ability to change. But again, it would be in and of the Creator and according to the *Grand Plan*. I would still be among the exceptionally privileged to have been a witness to the universe, and if life ends… it ends.

I cannot and will not advise you on religion, neither what you should think nor what you should believe. These are your choices and you have to decide how you move forward and find spiritual and emotional comfort in ways that matter and make sense to you. I chose to live my life deliberately, in awe of all the creations in the heavens and on earth. I got up from my knees begging for a life everlasting when I realized I had been ignoring the one that I was given. I knew most of my prayers were selfish in nature, insincere and many were hypocritical and I knew that my god knew it. I could no longer lie and try to deceive one that was all knowing. The Creator has never asked us to send up adoration and exaltation or to offer tribute in turn for answering our prayers; that is a requirement of our earth gods. And the earth gods were forced to make that bargain through the writings of our religious leaders, and the writings of religious leaders are the writings of men. According to those writings, man's purpose in life is to serve God. But the service of God must follow the rituals that they, and they alone set out, and none others. They teach that the earth and it's bounty as well as the universe was made by God for man; it was not! They teach that man was given a short term life on earth (a trial life) in order for him to prove that he is worthy of life everlasting in the heavens; that is of course, so long as he gives himself over to the faith and follows the dictates of the religious leaders. I can find no proof that God is holding auditions on earth for the best worshipers to join in the multitude of heavenly worshipers. Though, if you believe that you exist, then your existence is proof that The Creator gave us *"this specific life at this specific time"* along with *free-will* in order for us to choose our own path and purpose, not just to follow the dictates of some demagogue. If we were only put here to grovel on our knees and pray and worship at the foot of a manmade altar, there would be very little purpose in life and there'd be no need for us to have free-will. But, because man is far better at following than leading, we capitulated and willingly gave up our ability to choose and free-will for the security of the group; and that made us cowards and slaves to the cults of religions and the religious leaders.   And for that, we only have ourselves to blame.

We became complicit with our religious leaders and created the earth gods because we needed reassurance and protection from the forces of nature and from ourselves. We needed to know that we were not alone and on our own in such a vast and frightening world. And we needed a superior being to curtail the beast residing in man. Man's greatest fear is the fear of death and our gods and religions offers us *"hope"* of eternal life. But in doing so, they set out a set of rules that are designed to keep man humble and subordinate to the gods if he expects to achieve the promise of the heavenly paradise. And to that end, religions act as a deterrent to man and his brutality. Because, without the strong influence and oversite from the all-mighty gods and an authoritative and powerful religious doctrine man will always step in and fill the vacuum. Man is not god, but history has shown that when men make themselves gods without fear of oversight from an all-powerful deity, he (with precious few exceptions) will create *hell on earth*. There'll be dictatorial rule, unimaginable human suffering, war, and genocide. When man makes himself god, there's nothing to keep him centered and there are no limitations; he becomes *evil incarnate*, and, there is no end to the number of civilizations that have crumbled and have been brought to their knees due to man's own ignorance, arrogance, megalomania and vanity.

The religions we've created have been a benefit to man and has allowed us to contemplate a higher level of existence and has expanded our sense of morality and ethics. But at its root, religion is an organized political enterprise whose purpose is to expand its reach and rule by enlisting additional followers, acquirer additional wealth and territory, political influence, and physical powers through the soldiers of the faith. When there is religious conflict, strife and war, it is normally brought about by religious leaders in their efforts to gain greater powers and influence. Following this line of thinking, and by my assessment, religion has both social, moral and ethical benefits for man as well as some significant political drawbacks which are both draconian and tyrannical in nature that are specifically designed to keep man subordinate and compliant. And again, this is not an indictment of religions, but an observation of how they work.

In follow along with the basic theme of this essay…when we eventually move beyond the earth to Mars, into the asteroid belt, to the moons of the gas giants and beyond, how will our earth gods and our religions be able to follow? They are all rooted in earth and the mythologies of this planet, and without earth as a reference point they will become the myths of the elders and magical stories that are destine to be set aside or rejected by the second or third generation of those born off-planet. And in this new age of enlightenment, how will the religious leaders be able to expand the realm of the earth gods, or be able to create a new and different pantheon of gods that could be believed and still be able to keep man on his knees? Without doubt, I am certain that the religious leaders are busy trying to find ways. And, I have no doubt that religious leaders of all stripes as well as many of their followers will thoroughly reject and condemn what I've written here. Each will sight their holy books and scriptures in an attempt to nullify any of my thoughts and any thoughts that conflict with their own, in much the same way as they reject and condemn the teachings and rituals of other religions, religious sects, and the various creeds. So, by my thinking, my position (as well as any you might hold) is as valid as any of the over 4,000 religions that are estimated to be practiced around the world. I am in no way suggesting that I have any answers about anything. In fact, I can only speculate and have no proof or facts about any universal truths. But unless and until someone or some group can provide *"proof"* that "one divinity" (and there can only be "one") has possession of unquestionable (universal) *"religious truth"* and has prescribed a particular set of religious rituals that were ordained by *"The Creator"*, I will continue to find comfort and truth in my own beliefs and will not be seduced by hope and cowed by the cult of religion, mysticism, magic and fear.

So in conclusion, if you only took away one thought from this section, let it be that you should use common sense and exercise your free-will in choosing your spiritual life's path and continue to worship and believe in the ways that give you comfort. I am no different than you and still hold tightly to my earth god. My earth god provides me a degree of comfort and security because I am unable to imagine the magnificence of The Creator. But in so doing, I stay mindful that *"religion is a creation and a political tool of man, not of God, and The Creator of the Universe has never entered into any bargains with man and has never made a single demand for adoration, glorification, or ever expressed a desire to be worshiped. The Creator and The Creation as well as the Grand Design of the Universe are in no way altered or manipulated by man's religions or beliefs; they are matters of fact and are what they are. They are not, and never have been, matters that were subject to man's influence, neither by his prayer or tribute. I recognize that man must have hope in order to thrive and survive, and that man finds hope in his God. I also have hope, and like you, my hope comes from a belief in my God. But, I believe that any thoughts that we can manipulate The Creator with prayer and tribute would be, by my assessment, the pinnacle of hubris and I can think of no greater affront to our maker."* So believe what you want and find hope, along with spiritual and emotional comfort wherever you can, whether in your earth god, in The Creator, in both or in neither but I would avoid getting caught up in man's religious rituals and cults because they are purposely designed to garner both power and wealth by politically seducing, indoctrinating, manipulating, intimidating, exploiting, and enslaving all of their followers as well as all of mankind in the name of their specific and preferred god…*and their god only!*

Moving Beyond the Earth:

Moving beyond the earth is an area of discussion that is better left to the futurist, dreamers, and engineers. I can't predict how our sciences will evolve or how, and in what ways, we will have to evolve as a species in order to move off into the stars. However, there are a few areas that I will talk to because they appear to me to be apparent but may not have actually been considered by the average person in the population. For example, we'll have to develop safe and reliable propulsion systems that will allow us to travel "at" or "above" the speed of light. If we can't do that, we won't be going anywhere. We could still head off into space but it would take many tens of thousands of years before we would reach another destination. Additionally, we'll have to change our perception of what it would mean to be human because once we head out of the solar system, we will be cut off from the rest of the human genepool and our evolution will move along a different path than those left behind on earth. And there must be a realization for anyone making the journey that due to physiological, genealogical and biological changes, their descendants will likely never be able to return to earth; it will be a one way trip. Further, we'll need to find ways to incorporate and protect our accumulated knowledge so it's never lost or corrupted. Because we have limited intelligence, reliance on our libraries to retrieve accumulated and stored knowledge will be essential to our survival. Without that stored knowledge we'd be thrown back into the dark ages, and if we were traveling through space or were in a hostile environment, the prospects for our extinction would be all but assured. Additionally, we'll have to discover ways to create and maintain a suitable environment during the journey and a suitable ecosystem at the final destination; this may require us to have the ability to do planetary terraforming as well as the ability to modify our own genetics in order to survive the new and alien environment. And lastly, we'll have to decide on who'll get to go and who'll be left behind. I have decided that I won't try to discuss that here because it is fraught with moral, ethical, social, and political pitfalls that can and will be debated to ad nauseam when the time comes to decide the nationality of a crew member, their political affiliation, and their genetic makeup (their racial affiliation) on the exploratory vessels, on the colony ships, and on the generational ships.

Propulsion Systems:

The greatest challenge will be propulsion, developing technologies that would allow us to travel to nearby star systems without requiring generational ships; ships, which upon arrival at the destination disembark some level of descendants of those that initially embarked upon the journey. And because light speed is considered the universal speed limit and mathematically unattainable at this time, means we will have to develop new math and physics before interstellar travel can be made viable. Current technologies will allow us to colonize the solar system but even then, it would require multiyear voyages out and back. A visit to our closest neighboring star (using current state of the art propulsion) would take over 70,000 years each way.

For the sake of argument, let's imagine that we could find a way to travel at light speed. If that were possible, moving about our solar system would become more like booking a trans-Atlantic or Pacific flight of around 6 to 12 hours of flight duration. The ships that we would need for these relatively short hops would only have to have a similar capacity as some of our larger passenger aircraft and cargo planes. But even traveling at that speed it would still take over 2 years to reach the outer boundaries of our Oort cloud and another 2 years for us to be able to visit our neighboring star system. For those trips we would have to have much larger *Colony Ships*; ships that were equipped with artificial gravity and were self-sustaining with multiple redundant backups. In addition, no ship should ever make an interstellar journey alone. There should be no less than 4 fully independent ships making the same journey with a minimum spacing of 10 billion miles between them should any ship require support or rescue. This would be equally valid if we chose to colonize the galaxy. If that were our choice, we would have to build *Generational Ships*; ships that might need to be as large as small cities. It would be at this point where man will cease to be a terrestrial being and will become a galactic voyager. If we were able to build these light speed chips, traveling from the top of our galaxy to the bottom would still take 30,000 years, and traveling across the galaxy (from one side to the other) would take

120,000 years. So in essence, to be able to travel at light speed would be a good start for traveling around our solar system but it would be far too slow for interstellar travel and totally inadequate for galactic exploration. According to the latest count there are over 400 billion galaxies in the visible universe. I will leave any discussions of intergalactic travel to those that break the light speed barrier because they would then have to go on to develop propulsion technologies that were many multiples beyond light speed. But I will only say that even at light speed, it would still take 2.2 million years to travel to Andromeda (the closest galaxy to our own). We would obviously require new propulsion science to make the leap in to intergalactic travel; assuming it would ever be considered to be in the realm of possibility. Therefore, in view of the incredible distances involved and because we don't have a way to fold or bend space, overcoming the light speed barrier would be considered the first step and essential for us to get out to the handful of stars within 50 lightyears of earth. But to date, ***"nothing"*** can travel faster than the speed of light though gravity and a handful of subatomic particles (neutrinos) are known to approximate light speed[4]. If we ever found a way to overcome the light speed barrier, it would be a major achievement and a good first step that would be essential for our initial exploration of the galaxy. But in galactic terms, it would only be a small step forward. It would be similar to sitting on the porch of a house in a very large city. We could see the people; the cars and buses, trains and plains pass-by on their way to far-off locations. But we would only be able to wonder about those places because we would be limited to light speed and that prevents us from moving beyond the safety and security of the top step of

---

[4] I have 2 thoughts here, both of which appear to be exceptions to the light speed barrier. The first seems to be a contradiction of empirical observations, and the second is based upon a theory. Scientists have observed matter being drawn into black holes and have reported that the gravity well inside of a black hole is so intense that nothing can escape its pull, including light. That being the case, it would suggest that the pull of gravity in a black hole is stronger (faster) than the speed of light. So in this instance the speed of gravity has overcome the speed of light. Then second is based upon the Big Bang theory which suggests that the universe exploded into existence from nothing at all. It exploded from *"nothing"* into *"nothing"*. The explosion and subsequent inflation is believed to have been so energetic that the infant universe expanded at speeds far greater than the speed of light. Therefore, it would follow that The Static Nothing had to retreat and shrink away faster than the speed of light in order to accommodate The Nothing that was becoming something; becoming the infant universe. If this theory holds, then it would mean that the infant universe and The Nothing that was surrounding it, but not a part of it, were both exceeding the speed of light.

the porch. Our galaxy is enormous and our universe is inconceivably vast. In keeping with this scaled down model, other neighborhoods and distant places, for example; any of the other cities in our country and places like Japan, China, Australia, Europe and Africa as well as the great oceans of the world would be beyond our reach and imaginations. We would only be able to see the houses on our street and what was happening in our immediate neighborhood (our local group of stars) and would have no clue as to the marvels that were awaiting us further out in the galaxy, and absolutely no real way to get there within any reasonable timeframe.

However, with our current technologies, the probability that we will colonize Mars, the dark side of Mercury, some of the larger asteroids in the asteroid belt and some of the moons of the gas giants are nearly all but certain in the not too distant future, but interstellar travel, without cracking the light speed barrier, would be a major challenge and without it, traveling across those vast distances will not occur.

It was precisely because of these scientific observations (scientific facts) that caused me to look beyond my earth god and religious teachings to The Creator and The Creation in order to help me to find answers to the many questions I had about the concept of "God", the absurdities and inconsistencies of religions, my own existence, as well as man's place within this inconceivably vast universe.

Moving on…

Human Evolution and Gene Expression:

At some point, we'll have to really come to terms with some hard facts about human evolution and change. We'll have to recognize that once we close the hatch of a generational ship that is headed out of our solar system, and if that ship is not capable of traveling at or above light speed, we'll have to reassess our definition of what it means to be human. Because, once the hatch is closed, the occupants aboard will be cut off from the rest of the world's populations gene-pool and once in flight, cut off from earth's gravity, protective shield, and earth's environment. Even a voyage to the closest solar system to our own would take many tens of thousands of years. During which time the ship's inhabitants would be exposed to artificial gravity, cosmic radiation, virtual confinement and a limited amount of genetic material. Each of these exposures (particularly the limited genetic material) would be a variant to conditions on earth and would have a profound effect on the evolution of the ship's population. In essence, earth's population would continue to evolve and the inhabitants of the generational ship would evolve but the earths' and the ship's populations would be evolving separately and along two ever widening tracks. Ultimately, the two populations may deviate to such an extent that they would eventually become two separate and distinct species; they would no longer be able to successfully share genetic material. The same would be true if we launch generational ships on multi thousand year voyages in all directions in the galaxy, or if two ships traveled side-by-side for thousands of years without sharing genetic material. Different genes will express themselves at different times during man's evolution and through the process of *"Natural Selection"*; specific genes will dominate while others will become recessive or dormant. The genetic changes occurring in one population will be separate and apart from those happening in another. Therefore, the definition of what it means to be human will be highly subjective and truly population specific.

If we flush this out a bit further, we'll eventually come to the very real conclusion that if man survives in the cosmos; his survival will be due in large part to his ability to change and adapt his genetics to his environment. Even if we cracked the light speed barrier as much as tenfold, it would still take more than 3,000 years to travel from the top of our galactic plain to the bottom and over 12,000 years to travel from one side to the other[5]. Then there's always the time it would take to get back home. If 10 ships were sent to different parts of the galaxy and 10 ships returned after 40,000 years, there would be 11 variants of humans on the planet (those from returning ships and the population on earth). The question would then become, "which population deserves the right to be called human? Which will get subordinated to humanoid status, and which will be considered a completely different species and no longer human at all?

It is also possible to have one population that evolved while the other remains static. This particular outcome would be based upon *"Particle Physics"* and what happens with a particle approaching the speed of light. According to Einstein's theory on *"Special Relativity",* in addition to the particle becoming infinitely more massive, which then requires infinitely more energy in order to increase its speed, time slows down and space contracts as the particle approaches light speed. This suggests that once light speed was achieved time would *stop!* If this were in fact a reality, and time stopped for anyone traveling at light speed, then theoretically, a traveler aboard a light speed vessel could conceivably travel anywhere in the universe in 1 second of relative time (the way they would experience time aboard the vessel). The travelers aboard the ship could travel 100,000 light years across the galaxy, spend 1 second at the distant location, and travel back to the earth in a total of 3 seconds of relative time[6]. But on earth, time would not have been altered and when they returned they would have found that man would have

---

[5] Physicist and propulsion engineers are currently exploring the feasibility of creating a *Warp Bubble* around a vessel that would allow the vessel to travel many magnitudes above light speed. If it were possible, it would represent another step forward in space exploration, but only a small step with respect to exploring the vastness of our galaxy.

[6] I recognize that there are a number of problems with this example and with any vessel's attempt at achieving light speed, not the least of which is the problem of navigation. Specifically, how could a light speed vessel be controlled when, *theoretically,* the ship and its occupants would be frozen in time?

undergone 200,000 years of evolutionary change. So, once again, which group is human, those that stepped away for what they perceived to be 3 seconds or those that had an additional 200,000 years to evolve?

Natural selection is a wildly random process (virtually a crapshoot), though with it, man will continue to evolve as a single species so long as there is access to all the genetic material in the gene-pool. But when a smaller group is separated from the parent group, the smaller group will be cut off and no longer able to exchange genes with the larger group. And while they will continue to evolve, their evolution will take them down a different path. Even if we were to colonize Mars, the dark side of Mercury or our own moon for that matter, the colonist would immediately be subject to a completely different and foreign environment (reduced gravity, background radiation, extremes in temperature, an artificial atmosphere, total darkness with respect to Mercury, and a diminished gene-pool. The body would do whatever it could to adapt to the environmental changes but if the conditions persisted, the body would begin trying to adapt genetically. A Geneticist or Anthropologist could give some insight as to an approximate timeline, the time needed for evolutionary adaptations to begin to be expressed. But just adapting to the physical changes would make a visit to earth by a 3rd or 4th generation Martian physically problematic. They would have to be able to adapt to earth's much stronger gravitational field, increased solar radiation, increased atmospheric pressures, and they would have little to no defenses against earth's microbiological flora. Over time, genetic adaptations would allow man to better acclimate to his new environment, but his evolution would be divergent to the population on earth. Consequently, and very much by necessity, any journey into deep space would likely be a one way trip. There would be no incentives to return because human evolutionary deviations would have eroded our human connections and identities with respect to those left behind. Additionally, humans are fiercely tribal and barely tolerate people with different skin colors, those that speak a different language, those from a different race, nationality, culture, or religious group. So it is highly unlikely that we would ever be able to accept other humanoids that evolved along a different path then we did? If ever a generational ship returned to earth after wandering the galaxy for 50,000 years without finding as suitable planet or moon to colonize, how would they be received by earth's current inhabitants? Both groups would have a legitimate claim to inhabit the planet but after 50,000 years of separation, they would have absolutely nothing in common, not culture, nationality, politics, morals, rituals, language, religions, not even their appearance; *nothing!* Invariably, after 50,000 years, I would expect that one group would be more technologically advanced than the other. And because both groups would be sharing a common ancestry, both would be highly intelligent, cunning, unpredictable and predatory. The only question that I would have at this point, and giving a nod to Star Wars, is "during the quite predictable and unavoidable war, which group would represent *The Empire* and which would be considered "*The Rebel Alliance*"?

Just to recap: Man will continue to survive as a unique species on earth as long as the planet supports human life, or until such time as he annihilates himself. But if man begins to venture beyond the earth and begins to colonize the solar system or if we become *"deep space mariners"*, our definition of what it means to be human will have to change, because we, as a species, will change. In time, natural selection and population isolation will begin separating us into distinct and unique branches on the human evolutionary tree. However, natural selection has a dark side and some genetic mutations can be lethal. If any lethal genetic mutations were to appear and the population did not have the knowledge and ability to correct or eliminate them from the gene-pool; the entire population would be put at risk, become developmentally challenged or, far worse, the mutation could cause the population to go extinct. Therefore, there should be a process to rigorously screen everyone on the outbound voyage to assure that none test positive for any genetic markers that could threaten the viability of the long term mission and the lives of the ship's inhabitants. For example, the markers of note should be those that screen for mutations that are known to be lethal to the carrier or those that result in debilitating impairment; i.e., cancers and Alzheimer's respectively. In addition, it would be essential for each outbound ship to take with them as much of the known medical technical knowledge as was available in order to treat and/or eliminate toxic or detrimental genetic mutations should they appear. If left unchecked, over time, such mutations could spread through the ship's population like a plague, and that population could be lost.

This brings us to the importance of preserving and protecting our libraries and technologies; our ability to collect and secure our learned knowledge and having the ability to access them when needed and shield them from ever being corrupted or lost. I will talk to that next.

Accumulated Knowledge:

Arguably the ability to control fire, the invention of the wheel and writing are touted as major milestones in man's evolutionary development, But one of man's creations that rivals them all has been overlooked and has not been given the recognition that it so richly deserves, that creation is the library; the repository of learned knowledge[7]. Without libraries, our total knowledge would be limited to those things we were able to learn over a lifetime and that small percentage of which we were able to pass along to the next generation. Most of what you've learned and even more of what you were taught would be lost to successive generations. Additionally, without libraries, there would be no way to capture collective knowledge; knowledge acquired by those in different groups and subgroups, those having different experiences, or specializations. Everything would have to be learned, then once forgotten, would have to be re-learned, and re-learned over and over again. And this is exactly what happened throughout thousands of years where man made little technological advancement. There is no evidence that we are any smarter or more intelligent "today" than we were 50,000 years ago. I would expect that there are no remarkable intellectual or cognitive differences between the hunter gatherers, of that time, and the indigenous and isolated people of The Amazon and in parts of Africa today. And I would suggest that the differences between them and us can be distilled down to the availability of stored knowledge; libraries. It was the library that allowed us to begin taking greater and greater leaps forward and has brought us to the age of science, and for our purposes, space travel.

There have been any numbers of studies that have attempted to quantify intelligence as a product of race or sex. And without discrediting their findings and addressing their more basic political agendas, their findings show that 68% of the world's population has a *"Normal IQ"* of 85-115, and 95% of the world's population has an *"Average IQ"* of 70-

---

[7] For our purposes, a library is any discrete or collection of transcribed or recorded information regardless of the medium, whether written, recorded, or visual as a stand- alone item or as part of a physical collection or as an entire electronic database.

130. And while we can continue to argue who's smarter than whom, the mean value *(Mean IQ)* of man's intellect appears to cluster around 100. Our libraries have allowed us to study the works of our predecessors and to make small and incremental improvements on works that were started and further improved by others. We seldom come up with fresh, new ideas and inventions. Even when we do, they are born imperfect and require constant refinements. And if we were traveling through space aboard a generational ship without the benefit of our scientific, technical, and medical libraries we would have to reinvent everything that we needed to overcome every possible obstacle we would confront. I can't imagine such a journey ending in success because, as a species, we're just not that smart! So we'll have to rely on all our libraries to supplement our informational shortfalls and intellectual deficits. All of the ship's systems, navigation, environmental, life-support, propulsion, computerization, hydraulics and electronics, defenses, shielding, medical, and on-and-on would eventually fail, as would any of their back-up systems, during a 20-50 thousand year journey. And the loss of any one system could be catastrophic to those aboard if they lacked the knowledge to repair them and bring them back online. Additionally there would have to be a documented system of governance; standards of conduct, ethics & morality, and written laws in order to prevent chaos, mob rule and a failed mission. The voyagers would also need to have access to information on human genetics and terraforming if genetic modifications became necessary during the voyage or at the new location where terraforming might be required. Then there is the need to satisfy man's inquisitive nature and his desire to learn. Man will need the knowledge accumulated in his libraries to serve as a baseline upon which he can continue to challenge and improve upon the technologies, sciences, and the arts as well as the humanities as he strives to become a higher functioning being.

With the help of our libraries our journey to the stars would have a real possibility of success. The voyagers that disembarked at the new location would do so as a highly enlightened and technically advance people, bringing with them sophisticated libraries that would bear little resemblance to those at the start of their journey. Their libraries would have been greatly improved throughout their journey with new sciences, technological advances and strategic updates and refinements. As explorers, they would be far better prepared to meet any of the challenges they would be likely to face. Without the initial libraries or if the libraries were corrupted or lost during the journey, there would be no one to disembark because the ship and its occupants would have been lost along the way.

We are living in *"The Age of Information,"* where any and every scrap of information is only a click or a search away. We take our informational resources for granted failing to realize the disaster it would wrought on our entire civilization if our libraries were ever lost. And while the creation of libraries has not been considered a major leap forward in man's evolution as has fire, writing, and the wheel, without libraries we could have our wheels, writings, and our fires but we would not have even developed, *technologically*, so much as to the level of those that lived in the Dark Ages. Without question, we owe our current progress and our future development to the creation and continued maintenance of our collective libraries.

Moving on…

Environmental and Ecosystem Suitability:

I won't make an attempt to summarize the multitudes of sciences and technologies that would have to be perfected in order to assure the environmental safety of anyone leaving the protective envelope of earth for the hostel confines of space; the list would just be too long. I will also avoid any discussion of *cryostasis, hibernation, or suspended animation* for those making the journey because of the complicated nature of the subject. However, I will note a few problems with the processes. Cryostasis, hibernation, and suspended animation will not stop the destructive nature of time, and with time, there will be a deterioration of the voyagers' physical bodies as well as their intellectual capacity. The process itself could kill them. Additionally, if the voyagers slept for hundreds, thousands, or even a million years, there would be no intellectual or technological advancements made during that time and they could very well be overtaken and superseded by those embarking on the same journey thousands of years after their departure and would be thousands of years more advanced; both technologically and genetically.

Leaving cryostasis, hibernation, and suspended animation behind, we have taken some steps in the right direction and are learning more about what it will take to protect and preserve life in microhabitats. To date, we've made orbital flights, built and occupied orbital platforms such as Salyut, Skylab, Mir, and the International Space Station and we've landed on and briefly explored the surface of the Moon. Each required the development and installation of suitable environmental systems that were compatible with earth's environment. But these were all small and rather high maintenance systems designed to protect our astronauts and cosmonauts for a relative short term of exposure to the vacuum of space. They would be totally inadequate for long duration space flights or for the colonization of any moon or asteroid. Suitable long duration environmental habitats will require new approaches to life support systems and new thinking.

Working through technical and mechanical challenges will only require time and critical thinking. However, there will be a time when we will reach a point where major moral and ethical decisions will have to be made; a point of no return. This will be the instant we come to terms with the realization that environmental systems alone will not be enough to support human life for long duration flights; flights that would span many generations and many thousands of years. This will be when it becomes apparent that we'll have to make changes to our biology through genetic manipulation in order to sustain ourselves during the journey. When we take the decision to modify our genetic profile to improve our survival chances in space, we will do so by running head long into social, ethical, religious, and moral taboos. Tinkering with our genetics will likely be *"a bridge too far"* for many in the population, particularly those that are still guided by the religious writings and teachings of men. But, if we are to have any hope of making such journeys, we'll have to recognize and make the required adaptations ourselves and not put our fate in the slow and unpredictable process of natural selection. This then, would also suggest that those making the voyage would have undergone many of the genetic modifications for many generations long before disembarking from earth.

Upon reaching an extrasolar system planet or moon, the mariners may well be confronted with a second ethical and moral challenge, *"what to do with the indigenous life they find there"?* Through our observations of the universe, we've found a multitude of planets in orbits within the habitable zones of their stars[8]. And we refer to them as second earths. The implication is that those planets would be the most likely candidates to be able to support life; therefore, they would be the prime destination targets for human expansion into the galaxy. And if the planets are likely to support life, life may already have taken hold there. However, without the ability to travel at light speed, it would take over a million years to reach the nearest of such planets. Even then, they may find liquid water but the planet would be far from a utopia. The gasses in the atmosphere as well as the flora and fauna could be toxic to the new arrivals. In addition, the planet could be inhabited by species that clearly require that environment for survival. Further, there could be

---

[8] The habitable zone is the area around a star where water could exist on a planet or moon in a liquid state.

intelligent beings that appear antithetical to what the new arrivals understand to be intelligent life. Or, better yet, the indigenous life could react in a similar way as we would and become hostile toward the invading aliens. After a million years of travel and thousands of human generations that have come and gone, will ethics or morals even come in to play; will they even be considered? Would the mariners just turn around and head off on another million year journey hoping to reach another earth-like planet that's not already inhabited, or will they just commandeer this one? Sadly, we know the answer to this and it should tell us something about what we could expect if we were ever visited, to assume otherwise would be naïve and simpleminded.

The optimistic scenario would be to find a *completely dead planet* near the outer edge of the habitable zone where the water was still trapped in ice. Slowly warm the planet and start the process of terraforming its atmosphere and surface over many thousands of years from orbit and seed it with bacteria, flora, and fauna from earth until the planet became suitable for human habitation. This scenario would be ideal, but realistically, not probable because I believe that we are destined to find life wherever we go in the galaxy and most abundantly on earth-like planets. By necessity, terraforming any terrestrial body would require the complete sterilization of that body, followed by an introduction of new and alien species. And, terraforming of any world or moon where life already exist would be unethical, immoral, and an ***act of global genocide***.

Now, I'll leave it to you to try and decide the selection process; how would you decide who would get to go. How would you select individuals from earth's populations to become the genetic seed stock and heritage of humans as we move into interstellar space? Who would you select and who would be rejected, and why? Are you comfortable with the way you decided?

Now I will move on and talk my theory of dimensional space…

Dimensional Space:

There are a multitude of concepts that we are just not equipped to understand because they are beyond our intellectual ability to process. For example, the concept of nothing, the nature of God, the dimensions of space and time, unlimited intelligence, and dark matter and dark energy just to mention a few. And while each of these is clearly tangible and definable in the natural universe, they are well beyond our comprehension because they are beyond our experiences and imaginations. We are only sentient in three dimensions and can only experience time when it's measured in seconds, and we are blind to higher dimensions. We can't even point to the fourth dimension which would be at a right angle from the first, second, and third dimensions; just as they are at right angles to each other. What we can perceive and imagine is limited to our senses, how they interface with reality, and our inability to imagine what we cannot see and those things that appear, from our perspective, impossible. Each successive dimension strips away barriers to the hidden universe, and for our purposes, it would be that part of the universe that is visible at and above the fourth dimension. For example, try to imagine existing in a one dimensional universe. There would be the awareness of time, but our entire existence and perception of our universe would be reduced to whatever could be experienced by living in a flat dot. We would be able to move and visualize our environment from right and left, but not up or down or forward or backward. If it were possible to see beyond our dot, it might appear to us as though we were standing on some abandon train tracts in a thick fog and on a vast plane. We wouldn't be able to see anything above us or below. We wouldn't be able to see anything on either side of the tracks, just the tracks as they were laid out in the distance, appearing to taper and taper until they disappeared into the fog. If we added another dimension to the first, we would become privy to vastly more information. Though, our world would still be flat, we could move right and left and up and down but would still not be able to interpret what was really there. Try to imagine yourself moving around on a mile-wide jigsaw puzzle of 200,000 pieces. Each piece provides us valuable new and different information but you would still have no clue as to what the real picture was. The third dimension gives us depth and perspective. In the third dimension, we are no longer trapped as flatlanders and can move right and left, up and down, and now forward and back. We can see all the train tracks, the sidings, towns and cities surrounding the tracks, the sky and the earth that was oblivious to us in the first dimension. And when we back away from the jigsaw puzzle we can see the full picture and understand its meaning. We can move on three axes within space and time, and can have a greater appreciation of our environment…but only marginally. I say this because we really don't have any idea of what we're missing and are blind to everything beyond the third dimension. The additional information we need to understand our universe is hidden within other dimensions and is beyond our comprehension. For example, if we only existed in the first dimension we would not have been able to imagine a "Y" axis; further, if we only existed in the second dimension we would not have been able to predict a "Z" axis. Therefore, it's not surprising that we can't imagine what spatial axis might exist in the fourth dimension. But because each additional dimension provides us more accurate and reliable information, we'll need to theorize what an existence in those dimensions might be like in order to acquire a higher level of understanding (intellect). And if there really are 10 dimensions of space and one of time (as theorized in *String Theory*), I would say that we have a ways to go before a valid picture of our universe would begin to come in to focus (i.e. the concept of nothing, dark matter and dark energy, the nature of God, etc.).

Now, not wanting to sound too pessimistic, but we'll need to solve at least two really major problems before we can begin to develop an understanding of the universe, God, and our existence within the Grand Design. The universe is vast, and until we develop the intelligence and the techniques to manipulate time and space it will remain a vast ocean which we can never cross. The speed of light appears to be the universal maximum and it is far too slow for our purposes. I believe that there is a solution to the light speed barrier but I fear that it may be hidden from us in higher dimensions and completely out of our reach and imaginations. And so long as we remain trapped in a three dimensional existence, there isn't the slightest possibility of us experiencing an existence in higher dimensions, overcoming light speed, manipulating time, or fully understanding the universe before us. But here is the contradiction, and it might be somewhat difficult to grasp (and a little spooky). I will argue that we already have all the answers to all the questions in the universe. But the answers are hidden from us in locked boxes in our current dimension and the keys are only apparent in increasingly higher dimensions. For example, if our existence was limited to the first dimension, there would be a reason why we couldn't look up or down, or backward or forward. Those spacial dimensions wouldn't exist in the first dimension so we could only experience space to our right and left. And that would be because in the first dimension space only existed along the "X" axis. But with the addition of each higher dimension exponentially more of the universe was revealed as well as it secrets. Consider this; if we exist in three dimensions and in time, then we exist in the first dimension along the "X" axis, in the second dimension along the "X & Y" axes, as we simultaneously exist in the third dimension along the "X, Y, & Z" axes. These existences appear to be facts and apparent, therefore I have to accept them as truths. But if they are true, then why can't I experience my current existence in the first or second dimensions? Why is it that I can only experience an existence in the third dimension? Using primitive logic this would suggest that if we exist in the third dimension we would also have to exist in higher dimensions as well because there can be no fourth dimension without having the other three. And it further suggests that we would be uniquely and exclusively sentient in all dimensions simultaneously but with only an ability to communicate to higher dimensions and prevented from communicating to lower ones. It would also imply that the variant of us that would exist in the 10th dimension would have already answered all the questions that are beyond our ability to answer in our current three dimensional existences. Additionally, because all the dimensions of both space and time would be apparent and intuitive to them, there would be no more secrets in their universe. They would possess near unlimited intelligence, and to that end, would truly know the universe's *Grand Design*, the solution to light speed and time dilation, dark energy and dark matter, up to and including, the nature of God.

Now, without taking too large a leap of logic, I want to consider the universe, our existence within it, the Grand Design, and the nature of God using a different approach. It is my understanding from *The Standard Model of Cosmology* that the universe, the visible universe, the part that is visible to us, only accounts for less than 5% of what is actually there; and we are blind to the other 95%. It is also my understanding that about 27% of the invisible 95% is made up of *Dark Matter* and the other 68% is *Dark Energy*. I can either assume that the 95% of the universe that is invisible to us is merely the benign ether of the cosmos and has no real purpose, or I can speculate on what I believe could be its origin and purpose. I will of course, speculate on the latter and the concept of a multi-dimensional universe; *The Grand Design*.

The Grand Design...the Multi-dimensional Universe:

The notion that we live in a multi-dimensional universe is not a new one. And for the most part, the dimensions have been described as being just a fraction of time and space out of phase from the next. But each dimension has been depicted as being virtually identical but existing in a different spacial region and having a different reality and possibly with different laws of physics. I don't have the math to argue for or against the multi-dimensional theory, but because we exist in three dimensions, I can't argue against the possibility that there may be more. Accordingly, if I accept the theoretical physics presented by the *String and Membrane Theorist*, there would be the possibility of 10 spacial dimensions plus one of time. And by my thinking, I've found that it meshes perfectly to my perception of the physical universe and its design with respect to time and space.

In order to describe my theory of the multi-dimensional universe I am forced to use a primitive prop to describe abstract concepts and will ask you to use your imaginations to follow. But as you do, please note that what follows are *my assertions, my argument, and a grand proof of nothing!*

Imagine our solar system containing 10 planets in concentric orbits surrounding our sun. I am using the solar system to represent our entire universe, and the Sun to represent the instant of the *Big Bang*, and the planets to represent the dimensions of time and space. If time and space moved out as the universe expanded behind them, there would be nothing ahead of them and nothing to see. The physical and visible universe would only exist inside of the bubble of space-time. Therefore in my model, you can only look inward (toward the Sun) because nothing exist outside the bubble and the planets have not yet begun to coalesce. The Sun would be dimensionless, just turbulence and chaos. Anomalies in time and space would have created the planet Mercury, the first dimension of time and space. Hence, the visible universe would be limited to the "X" axis and what could be viewed from Mercury looking along the horizontal. Venus would introduce the second dimension (the "Y" axis) and the universe would consist of a flat view of the Sun and the planet Mercury. Earth is located at the third dimension that provides depth and perspective with respect to the first and second dimensions (the "Z" axis). Each successive dimension incorporates all the lower dimensions and the visible universe becomes exponentially larger and more and more apparent and intuitive with each successive dimension. Problems arise when we try to point to Mars (the fourth dimension of time and space). It is outside the orbit of earth as well as our imaginations; its visualization is impossible. And if only 5% of the universe is visible to us, it's because of our limited perspective in the third dimension. The entire universe is there in the other 95% but it is invisible to us because we can't turn around and look behind us; we can only see those things that are inside the orbit of the earth; the Sun, Mercury, and Venus. So because we don't know what's there, we've chosen to refer to it as dark matter and energy. It is only when you reach the tenth planet (the tenth dimension) is the universe revealed showing all the dimensions of space and time. With each successive dimension more information is revealed and with more information there is greater and greater intelligence. So…what is intelligence, where does it come from? And if we have an existence in the tenth dimension and would possibly have near unlimited intelligence, why couldn't or wouldn't we communicate to ourselves in the third? I have no answers here either. But I believe that there is a *life force* in the universe, and everything that has life, by any definition of life, has some degree of intelligence; whether cognitive, instinctual, or sensory. And they all (regardless of dimension in space-time) provide information up to their highest level of existence, to the tenth dimension or higher. So now the question becomes "why", why is all the information transferred to our highest level of existence? Is the universe alive? Better yet, is the question obtuse because it is being posed by someone living in the lower dimensions of existence and this is the best question I could put forth? If there are 7 higher dimensions above the third then my question just might be irrelevant, meaningless, and naive in the tenth dimension. In the tenth dimension there could be no life or death as we have defined it, but a higher level of spacial existence, intellect, and the passage of time. I would argue that if we exist in the third dimension (in some form) we would also have to exist in some form in the tenth. And with respect to communicating from the tenth dimension down to the third, the information isn't needed in the third dimension and would likely not be understood; as a consequence, it was not part of the universe's grand design. But a couple of additional questions beckon. The first is, *"where are the higher dimensions of space-time located"?* As each succeeding dimension encapsulates all the lower dimensions and expands upon space-time revealing more of the universe, the question becomes *"are the higher dimensions and our existence in them still earthbound or are they spacial and ethereal without the need of our planet or for us to continue in a three dimensional physical form"?* And the second is *"once a copy of **all of space-time** has been recorded in a lower dimension and transferred to a higher one, does the lower dimension continue to have relevance or does it become superfluous and collapses out of existence"?* Once again, I don't know.

However, what I do know is that it's very easy to write yourself into a ditch or a dead-end where there is no possible way out. And even when you can see it coming, often times, there is little that you can do to prevent it. This appears to be one of those times. So with eyes wide, I can see at least two conundrums coming.

Just for the moment, let suppose that the *Multiple Big Bang Theorists* are right, and the universe repeats a cycle of expansion (bang) and contraction (crunch) on a regular basis. If the multidimensional design of the universe is correct, what becomes of the information collected and recorded in the highest level of intellect; the tenth dimension?   If the information is lost during the subsequent contraction the repeated banging and crunching would serve no purpose, alternatively, if the information is preserved, then how? If the initial expansion of the universe was due to the *Big Bang*, and the subsequent expansion in to nothing, creating time and space was caused by the inflation (stretching) of space-time to accommodate the various dimensions of space, then the universe would stop its expansion once the highest dimension has been reached.   At that point, the forces of gravity in each dimension would begin to collapse the dimensions of space-time resulting in another *Big Crunch*. All the information collected during the evolution of that universe and any reference to it would be lost if there was no way to transfer the information to an outside repository. But in a closed universe, nothing can enter, and nothing can leave. But it just might be possible for The Nothing, forming the bubble around the universe, to make a copy of the information in the highest dimension. Thereby, the information would be conserved before the collapsing universe crushes itself out of existence and rebounding into a new and different universe with different physics and realities. Let's go a step further, if the information can be copied, why not the tenth dimensional variant of us?

Now to the first conundrum…, "where is The Nothing, how and why would the information be stored there, and who or what would need or want to have access to it"? The Creator would have no need for the information because it is all-knowing, but a lesser entity might; for example, if copies of us were recorded and exist alongside the information in The Nothing, we would be the ones that would be in need of the information and we would become the "Intellect" of The Nothing. This would clearly suggest that there might be an existence outside of the universe or at the very least, that a record of this and pass universes have been recorded and cataloged[9]. If that were actually the case, it would further suggest that we, or a record of our existence, would become *eternal*. A further complication and just to make things a lot more confusing is the possibility that we could no longer be alive and living in our universe at all, but exist as part of a record of our pass universe. If that were true, it would give credence to the philosophy of *"Predeterminism"* (predestination), where everything we do had already been done; nothing could be changed or altered and everything would have already been fated and preordained.

---

[9] My suggestion here is that there is a possibility of an *"Eleventh Dimension"*, a dimension of space-time that exists completely outside of the universe and is independent of it. Or, the eleventh dimension could be a reflection of the tenth dimension in The Nothing that is encapsulating the universe.

Moving on to the second…, earlier I assumed that there was a *"Life Force"* within the universe; the life force in the new universe would have no mission or perception of itself without intelligence. But, in the new universe, the life force and every other living entity appears to be operating with some degree of intelligence and appears to have been doing so at the very earliest stages of the universe's development. How, if everything in the prior universe was crushed out of existence during its collapse (including intelligence), where does the intelligence come from in the new one? Let's imagine that there is only space-time expanding against nothing at the instant of the Big Bang, before the creation of the first dimension. Some form of intelligence would have to be present and acting on space-time in order to create the first and subsequent higher dimensions. So where would the intelligence come from? I can only address this second conundrum using *default logic*. If you can agree that The Creator and The Creation is omnipresent and possesses unlimited intelligence. Then it would follow that The Creator and The Creation is all that ever was, all that is, and all that ever will be. This would include The Nothing surrounding the current universe and pass universes, and universes that are yet to be created. It also would mean that because The Creator possesses unlimited intelligence that intelligence would be in all places simultaneously (both inside and outside the universe), thus, the new and expanding universe would have to contain intelligence as well, as it is part of The Creator and The Creation[10].

Before any of you accepts my argument or rejects it out-of-hand, remember that I am attempting to fit together a ten dimensional puzzle that I can't see or understand from my three dimensional existence and totally inadequate intellect. Subjectively and objectively I am attempting to define the origins of intelligence in the early universe while speculating on the conservation of information during the final moments of our universe's existence. What would be your thoughts with respect to the universe's design? The world would welcome your imaginations and arguments as well as your predictions of how the

---

[10] This then would not make God a separate and distinct being, but God being the universe itself.

impossible would become possible at higher dimensions of space-time? You should consider making your assumptions public.

Now, a couple of follow-along questions require some attention. Specifically, "what is intelligence, and can intelligence operating over time be considered a proof or a definition of life"?

Intelligence and a Proof of Life:

There are basic physical laws governing the evolution and devolution of the universe and they are expressed in the language of mathematics. Because the laws cannot be violated and appear to determine the limits of physical reality, could they also represent the foundation and scaffolding for universal logic and intelligence? The laws of physics are static and locked in place but intelligence is a dynamic process, not an event, and is expressed over time without the use of mathematics. Intelligence requires an ability to be able to process information over time regardless of duration from Planck time to deep time and anything in between. And the processing of the information could take any form, from cognitive, instinctual, sensory, or chemically initiated but time is required in order to allow for the process and a reaction or to decide. Furthermore, if there is intelligence, by my assessment, there's life. Of course, there is a significant difference between something being alive and something being sentient and self-aware. But I have no idea as to where that transition occurs, but it is clear that intelligence would drive the process.

I think I should be clear here before I run the risk of being misunderstood, I am in no way suggesting that a computer database is either intelligent or has life. By my thinking, computers process inputted information into a system of algorithms, using a binary system of ones and zeros. They don't think, but output results that are based upon probability and the limits of their software. I recognize that there are arguments on both sides of this question and as computers become more and more complex and sophisticated, as well as imbedded in our lives and bodies, the arguments that computers being intelligent will only grow louder as will the question as to whether they become sentient and by default, alive. But at this point, I will take the position that to date, computers are not alive. The argument could be taken up again at some point in the future when *"quantum computing"* has been realized and perfected, but for now the argument has no merit. In addition, with respect to divine intelligence or intelligence without limit, they're beyond my comprehension and imagination so I will leave that philosophical discussion to the philosophers and theologians.

When I consider reality, I stay mindful that I'm not getting the full picture from my three dimensional prospective of the world. Intelligence could be everywhere and in everything in the universe and the universe itself, is likely (for lack of a more accurate description) alive. But I will confine my discussion to my own reality in an attempt to describe an invisible and intangible force or power that transcends our physical environment that we can easily describe and quantify, but which consistently escapes our best efforts to define; the essence and origin of intelligence itself. All of the publications I've found that purport to define intelligence, come up short by only putting forward a description of intelligence, not a definition of it. For example, would you consider the following a definition or a description of water. "Water is a liquid between 32 and 212 °F, a solid below 32 °F and a gas above 212 °F. It has a specific gravity of "1" and is wet and transparent in its liquid state". Would you consider this a definition or a description of water? Would a better definition be, "water is an inorganic bipolar molecule composed of 2 hydrogen atoms and 1 of oxygen"? The multiple verbiages put forth in an attempt to define intelligence are as equally descriptive and vague as the first for water. Examples would include; Intelligence is the ability to reason, to think abstractly, to learn, to apply knowledge, have situational awareness, and to manipulate one's environment, etc.,. I've found them all very good descriptors of what intelligence allows, but absolutely nothing about what intelligence actually is. If intelligence was easily definable there would be a single definition for it instead of multiple descriptors. And if I knew what intelligence was, I would tell you. But needless to say, I don't.

We know that the senses and a normally functioning brain or sensory receptors are essential to intelligence and environmental awareness. We know that the most basic single-cell organisms operate with some degree of environmental intelligence coupled with genetic instructions that allows it to take advantage of favorable conditions to feed, respire, and multiply. When conditions become hostile some organisms die out but some others have the ability to react by becoming dormant or by forming spores until conditions improve. These are not random events that are driven by physics, but by chemistry; they are universal and appear to be driven by some level of intelligence. Multicellular organisms (plants, animals, and insects) exhibit similar chemical responses to environmental changes, as well as sensory and spacial inputs. And when it comes to humans and many other higher animals we have the added ability to think and problem solve. But again, this is only a description of what intelligence allows, not a definition of what it is. Therefore, without a compelling argument of my own, I'm going to defend an argument that Socrates put forth over 2,500 years ago. In essence he argued that the truth about all things resided within each of us and not in the stars, in our traditions, in our religious books, or in the opinions of the masses. We are born with the knowledge of truth and more and more truth is revealed to us through the process of living, social interactions, and relentless critical self-examinations. If Socrates were right, then it would suggest that intelligence is not something that we acquire; it is a product of our existence and is fundamental to the life force. And by my thinking, life would cease to be a force within the universe without divine intelligence. Intelligence likely exists in its entirety everywhere but the capacity of any living entity to realize it intuitively is subject to their level of complexity and their ability to comprehend. Single-cell organisms have life and a single-cell intellect, and multicellular organisms are far more complex with a capacity to realize much higher levels of intelligence than the single-cell. As humans, we have far greater intellectual capacity and have evolved to such a high level of complexity in our three dimensional existence that we've become self-aware and identify ourselves as unique and individual beings. As a consequence of our individualism and intellectual narcissism, we've taken a further step in setting ourselves above and apart from all others. We've convinced ourselves that we are special and even favored above all other living things and that our self-aware life force, uniqueness and distinct intellect is proof that *God created us with an immortal soul*. But I believe that our assertions are a bit naive and arrogant because even with our increased intellectual capacity, we are still limited and are clearly at some great distance from any ability to acquire all the intelligence that may currently be available, but not apparent to us, in our three dimensional existence. The third dimension may only be *translucent* to us and not *transparent*, and we may just not have the capacity to comprehend everything that's really there; those things that should be obvious and glaring. And because of our cerebral inadequacies, we are no different than a mouse being exposed to the principles of *Newtonian Physics*. It wouldn't matter if it were the smartest mouse ever; the concept of physics and mathematics would just be too far beyond its mental abilities and every attempt at learning would be destine to failure. The only hope for the mouse would be to have the time to evolve to a higher intellectual capacity before it would be able to comprehend what was actually in front of it; the same would be true for us in our quest to understand all that should be apparent in our world. In addition, we would have absolutely no possibility of comprehending intellect that we could be awash with but would only be intuitive to those that would exist at higher dimensions of space-time and at higher dimensional intelligence; those that might be more favored than we presume we are.

Subsequently, in support of my Socratic argument, and with respect to my failed attempt to actually *define* intelligence, I will *describe* it as being **"limitless clarity without reasoning"** as well as **the intangible animator of the universal life force**; it is **the consciousness of the universe itself**. In addition, I would argue that **wherever there is intelligence, there is life**. And lastly, because you are approaching the end of this essay, it should not come as a surprise to you that because I believe that our universe is in The Creation and because The Creation is *in and of* The Creator, I believe that our Creator is alive. Therefore, and by my thinking, if our Creator is alive, it would follow that **our universe would also be alive!**

One final thought about the universe and intelligence before I close…, though it may seem to you to be unlikely and bizarre, it is not impossible for this universe to be *"your own special universe"*. To my knowledge, there is nothing preventing the entire universe from being a product of intelligence and imagination; a product *"of your own intelligence"* and yes, *"of your own imagination"*. And while we believe the universe to be billions of years old, it may have just sprung into existence at the instant that you became sentient and aware. It became what you imagined it to be and if that were the case, it will cease to exist at the instant of your death or when you become unresponsive and oblivious to the passage of time. In this reality, if the question were ever asked, "if a tree fell in the forest and no one was around to hear, would it have made a sound"? The answer would be *"No"!* In this reality, nothing would exist beyond your senses and if you were not there, there would be no tree to fall, therefore there would be no sound and there would not even be a forest. You would be the "Alpha" and "Omega" of this universe and reality would only exist where you exist and where you imagine. You have all at one time or another, suspected that this scenario could be an actual possibility and neither I, nor anyone else, can realistically prove it to be otherwise. The really big downside here is that if this were a reality, then every event that occurred over the course of your lifetime and every event recorded in history (both the renowned and the infamous) *"would have been your doing; it all would have been your fault",* and there'd be no one else you could blame! A further downside for the rest of us is at the instant of your death, we will also cease to exist; you will take your entire universe and all of us into oblivion along with you.

I will close by briefly summing up my thoughts from the Oort cloud below, but if any of this has caused you to revisit your own thoughts on any of the topics covered or prompted you to consider other possible theories, then I would consider my efforts here a success.

Summation:

Looking out from our Oort cloud I find that I can only be confident of two apparent facts; the first is ostensibly obvious; that at the time of writing this essay, *I existed*. And the second, if I existed during that time, then there had to be a force or power that caused me to exist; that force or power (regardless of its origin) would have had to have been *The Creator*. So it would follow that my existence and yours, is a proof of *The Creator…, a proof of God!* Everything else, objectively and subjectively, is open to further debate and analysis. For example, space travel, our place within the universe, the origin of the universe and its ultimate demise, dimensional space, God and the concept of religion, intelligence and life, and so on. In addition, I am well aware that my approach in putting forward some of my own thoughts and my own theories only added to the debate and may be counter to the endless theories that have already gained some level of societal as well as scientific acceptance. But my thoughts are my own and I believe they have some value; they have allowed me to approach these theories from a slightly different perspective. Specifically, I don't believe that life is an anomaly but a fixture in the universe, and I believe that there is a universal intelligence animating and driving that life force. I also believe that if there are higher dimensions of space-time, we would have an existence in each of them and a copy of our existence might be recorded and stored outside our universe and that copy would be eternal. Then, with respect to God and religion; I can only suggest that you believe how you choose and in ways that bring you comfort, but in my world, a belief in God does not compel a belief in a religion. Your positions on any and all of these topics may be in conflict to mine, and if they are your positions, than they have merit and should be vigorously defended. But if the positions you hold are from humanities' echo chamber and from the voices and noise of the crowd, then your positions will be hollow and difficult to defend because you will have doubt and won't be able to defend what you don't truly believe.

And with that, I conclude my thoughts from inside the Oort cloud.